IFIP Advances in Information and Communication Technology   427

Editor-in-Chief

A. Joe Turner, Seneca, SC, USA

Editorial Board

Foundations of Computer Science
  Jacques Sakarovitch, Télécom ParisTech, France
Software: Theory and Practice
  Michael Goedicke, University of Duisburg-Essen, Germany
Education
  Arthur Tatnall, Victoria University, Melbourne, Australia
Information Technology Applications
  Erich J. Neuhold, University of Vienna, Austria
Communication Systems
  Aiko Pras, University of Twente, Enschede, The Netherlands
System Modeling and Optimization
  Fredi Tröltzsch, TU Berlin, Germany
Information Systems
  Jan Pries-Heje, Roskilde University, Denmark
ICT and Society
  Diane Whitehouse, The Castlegate Consultancy, Malton, UK
Computer Systems Technology
  Ricardo Reis, Federal University of Rio Grande do Sul, Porto Alegre, Brazil
Security and Privacy Protection in Information Processing Systems
  Yuko Murayama, Iwate Prefectural University, Japan
Artificial Intelligence
  Tharam Dillon, Curtin University, Bentley, Australia
Human-Computer Interaction
  Jan Gulliksen, KTH Royal Institute of Technology, Stockholm, Sweden
Entertainment Computing
  Matthias Rauterberg, Eindhoven University of Technology, The Netherlands

IFIP – The International Federation for Information Processing

IFIP was founded in 1960 under the auspices of UNESCO, following the First World Computer Congress held in Paris the previous year. An umbrella organization for societies working in information processing, IFIP's aim is two-fold: to support information processing within its member countries and to encourage technology transfer to developing nations. As its mission statement clearly states,

> IFIP's mission is to be the leading, truly international, apolitical organization which encourages and assists in the development, exploitation and application of information technology for the benefit of all people.

IFIP is a non-profitmaking organization, run almost solely by 2500 volunteers. It operates through a number of technical committees, which organize events and publications. IFIP's events range from an international congress to local seminars, but the most important are:

- The IFIP World Computer Congress, held every second year;
- Open conferences;
- Working conferences.

The flagship event is the IFIP World Computer Congress, at which both invited and contributed papers are presented. Contributed papers are rigorously refereed and the rejection rate is high.

As with the Congress, participation in the open conferences is open to all and papers may be invited or submitted. Again, submitted papers are stringently refereed.

The working conferences are structured differently. They are usually run by a working group and attendance is small and by invitation only. Their purpose is to create an atmosphere conducive to innovation and development. Refereeing is also rigorous and papers are subjected to extensive group discussion.

Publications arising from IFIP events vary. The papers presented at the IFIP World Computer Congress and at open conferences are published as conference proceedings, while the results of the working conferences are often published as collections of selected and edited papers.

Any national society whose primary activity is about information processing may apply to become a full member of IFIP, although full membership is restricted to one society per country. Full members are entitled to vote at the annual General Assembly, National societies preferring a less committed involvement may apply for associate or corresponding membership. Associate members enjoy the same benefits as full members, but without voting rights. Corresponding members are not represented in IFIP bodies. Affiliated membership is open to non-national societies, and individual and honorary membership schemes are also offered.

Luis Corral Alberto Sillitti
Giancarlo Succi Jelena Vlasenko
Anthony I. Wasserman (Eds.)

Open Source Software: Mobile Open Source Technologies

10th IFIP WG 2.13 International Conference
on Open Source Systems, OSS 2014
San José, Costa Rica, May 6-9, 2014
Proceedings

Springer

Volume Editors

Luis Corral
Alberto Sillitti
Giancarlo Succi
Jelena Vlasenko
Free University of Bozen/Bolzano
Piazza Domenicani 3, 39100 Bolzano, Italy
E-mail: {luis.corral, asillitti, gsucci, jelena.vlasenko}@unibz.it

Anthony I. Wasserman
Carnegie Mellon University Silicon Valley
Moffett Field, CA 94035, USA
E-mail: tony.wasserman@sv.cmu.edu

ISSN 1868-4238   e-ISSN 1868-422X
ISBN 978-3-662-51534-1   e-ISBN 978-3-642-55128-4
DOI 10.1007/978-3-642-55128-4
Springer Heidelberg New York Dordrecht London

© IFIP International Federation for Information Processing 2014
Softcover reprint of the hardcover 1st edition 2014
This work is subject to copyright. All rights are reserved by the Publisher, whether the whole or part of the material is concerned, specifically the rights of translation, reprinting, reuse of illustrations, recitation, broadcasting, reproduction on microfilms or in any other physical way, and transmission or information storage and retrieval, electronic adaptation, computer software, or by similar or dissimilar methodology now known or hereafter developed. Exempted from this legal reservation are brief excerpts in connection with reviews or scholarly analysis or material supplied specifically for the purpose of being entered and executed on a computer system, for exclusive use by the purchaser of the work. Duplication of this publication or parts thereof is permitted only under the provisions of the Copyright Law of the Publisher's location, in ist current version, and permission for use must always be obtained from Springer. Permissions for use may be obtained through RightsLink at the Copyright Clearance Center. Violations are liable to prosecution under the respective Copyright Law.
The use of general descriptive names, registered names, trademarks, service marks, etc. in this publication does not imply, even in the absence of a specific statement, that such names are exempt from the relevant protective laws and regulations and therefore free for general use.
While the advice and information in this book are believed to be true and accurate at the date of publication, neither the authors nor the editors nor the publisher can accept any legal responsibility for any errors or omissions that may be made. The publisher makes no warranty, express or implied, with respect to the material contained herein.

Typesetting: Camera-ready by author, data conversion by Scientific Publishing Services, Chennai, India

Printed on acid-free paper

Springer is part of Springer Science+Business Media (www.springer.com)

Preface

The International Conference on Open Source Systems (OSS) celebrated 10 years of interchange, discussion, progress, and openness. This year we received a relevant number of submission and workshop proposals, on all aspects of open source. Moreover, the conference was honored with the presence of distinguished speakers from leading technology companies who shared their experience on how to achieve success on top of open source technologies. A total of 37 high-quality contributions from 18 countries were accepted in the final program, thanks to the effort and coordination of the program chairs and the rigorous reviews of the Program Committee.

In 2014, OSS traveled to Latin America, landing in the beautiful city of San Jose, capital of the lively Republic of Costa Rica. OSS 2014 took place in Latin America, a region that experiences an enthusiastic and continuous growth in the development and implementation of open source technologies. The high participation of Latin American authors in the main program of the conference was remarkable, as was the celebration of the First Latin Colloquium on Open Source Software. This colloquium brought together authors and audience members speaking diverse Latin languages, such as Catalan, French, Italian, Portuguese, Romanian, Spanish, etc.

OSS 2014 presented as a featured topic "Mobile Open Source Technologies." Given the widespread diffusion of mobile devices and the relevance of open source operating systems for such devices (not only Android, but also Sailfish, Ubuntu, and others) the 10th edition of the conference solicited in particular submissions that focus on this area of open source, targeting their architectures, design choices, programming languages, etc.

This volume presents the 37 peer-reviewed contributions that made up the main program of OSS 2014. We organized the contents in eight parts, each concerning a relevant area of open source: open source visualization and reporting; open source in business modeling; open source in mobile and Web technologies; open source in education and research; development processes of open source products; testing and assurance in open source projects; global impact on open source communities and development; and case studies and demonstrations of open source projects. We heartily believe that this book will provide useful insights into the current state of the art and the practice of open source software development, technologies, project management, education, and applications.

For one decade, the OSS series of conferences has been a key reference to the open source research and practitioner community. This would have not been possible without all software developers, authors, reviewers, sponsors, keynote speakers, organization committees, local liaisons and all the involved staff who generously contributed to make OSS successful since its very first edition. We would like to gratefully thank them all for their interest, time, and passion.

March 2014

Luis Corral
Alberto Sillitti
Giancarlo Succi
Jelena Vlasenko
Anthony I. Wasserman

Organization

General Chair
Anthony I. Wasserman

Program Chair
Giancarlo Succi

Organizing Chair
Jelena Vlasenko

Proceedings, Web, and Social Media Chair
Luis Corral

Tutorial Chair
Stephane Ribas

Workshop Chair
Greg Madey

Panel Chair
Bjorn Lundell

PhD Symposium Chair
Andrea Janes

Experience Report Chairs
Mauricio Aniche
Carlos Denner Santos Jr.

New Ideas Track Chair

Alessandro Sarcia

Poster Chair

Kathryn Ambrose Sereno

Mobile Track Chairs

Mikko Terho
Ignacio Trejos

Industry Liaison Chair

Alessandro Garibbo

Financial Chair

Alberto Sillitti

Local Organization Chair

Otto Chinchilla

Program Committee

Wasif Afzal
Chintan Amrit
Claudia P. Ayala
Mauricio Aniche
Luciano Baresi
Cornelia Boldyreff
Andrea Capiluppi
Otto Chinchilla
Paolo Ciancarini
Reidar Conradi
Luis Corral
Francesco Di Cerbo
Carlos Denner Santos Jr.
U. Yeliz Eseryel
Jonas Gamalielsson
Carlo Ghezzi
Jesus M. Gonzalez-Barahona
Imed Hammouda
Scott Hissam
Netta Iivari
Paola Inverardi
Andrea Janes
Stefan Koch
Fabio Kon
Luigi Lavazza
Bjorn Lundell
Gregory Madey
Eda Marchetti

Sandro Morasca
Juan Ramon Moreno
John Noll
Witold Pedrycz
Etiel Petrinja
Mauro Pezze'
Rafael Prikladnicki
Stephane Ribas
Dirk Riehle
Gregorio Robles
Francesco Rogo
Daniela S. Cruzes
Denis Roberto Salazar
Alessandro Sarcia'
Walt Scacchi
Charles Schweik

Maha Shaikh
Emad Shihab
Alberto Sillitti
Diomidis Spinellis
Megan Squire
Klaas-Jan Stol
Eleni Stroulia
Marcos Sfair Sunye
Giancarlo Succi
Mikko Terho
Davide Tosi
Ignacio Trejos
Aaron Visaggio
Jelena Vlasenko
Eugenio Zimeo

Table of Contents

Open Source Visualization and Reporting

Code Review Analytics: WebKit as Case Study 1
    Jesús M. González-Barahona, Daniel Izquierdo-Cortázar,
    Gregorio Robles, and Mario Gallegos

Navigation Support in Evolving Open-Source Communities
by a Web-Based Dashboard 11
    Anna Hannemann, Kristjan Liiva, and Ralf Klamma

Who Contributes to What? Exploring Hidden Relationships between
FLOSS Projects 21
    M.M. Mahbubul Syeed and Imed Hammouda

How Do Social Interaction Networks Influence Peer Impressions
Formation? A Case Study 31
    Amiangshu Bosu and Jeffrey C. Carver

Drawing the Big Picture: Temporal Visualization of Dynamic
Collaboration Graphs of OSS Software Forks 41
    Amir Azarbakht and Carlos Jensen

Open Source in Business Modeling

Analyzing the Relationship between the License of Packages and Their
Files in Free and Open Source Software 51
    Yuki Manabe, Daniel M. German, and Katsuro Inoue

Adapting SCRUM to the Italian Army: Methods and (Open) Tools 61
    Franco Raffaele Cotugno and Angelo Messina

Applying the Submission Multiple Tier (SMT) Matrix to Detect Impact
on Developer Interest on Open Source Project Survivability 70
    Bee Bee Chua

FOSS Service Management and Incidences 76
    Susana Sánchez Ortiz and Alfredo Pérez Benitez

Open-Source Software Entrepreneurial Business Modelling 80
    Jose Teixeira and Joni Salminen

Open Source in Mobile and Web Technologies

Towards Understanding of Structural Attributes of Web APIs Using
Metrics Based on API Call Responses 83
   Andrea Janes, Tadas Remencius, Alberto Sillitti, and Giancarlo Succi

Open Source Mobile Virtual Machines: An Energy Assessment
of Dalvik vs. ART 93
   Anton B. Georgiev, Alberto Sillitti, and Giancarlo Succi

Improving Mozilla's In-App Payment Platform 103
   Ewa Janczukowicz, Ahmed Bouabdallah, Arnaud Braud,
   Gaël Fromentoux, and Jean-Marie Bonnin

A Performance Analysis of Wireless Mesh Networks Implementations
Based on Open Source Software..................................... 107
   Iván Armuelles Voinov, Aidelen Chung Cedeño,
   Joaquín Chung, and Grace González

Use of Open Software Tools for Data Offloading Techniques Analysis
on Mobile Networks 111
   José M. Koo, Juan P. Espino, Iván Armuelles, and Rubén Villarreal

Open Source in Education and Research

Crafting a Systematic Literature Review on Open-Source Platforms 113
   Jose Teixeira and Abayomi Baiyere

Considerations Regarding the Creation of a Post-graduate Master's
Degree in Free Software... 123
   Sergio Raúl Montes León, Gregorio Robles,
   Jesús M. González-Barahona, and Luis E. Sánchez C.

Lessons Learned from Teaching Open Source Software Development 133
   Becka Morgan and Carlos Jensen

A Successful OSS Adaptation and Integration in an e-Learning
Platform: TEC Digital... 143
   Mario Chacon-Rivas and Cesar Garita

Smart TV with Free Technologies in Support of Teaching-Learning
Process 147
   Eugenio Rosales Rosa, Abel Alfonso Fírvida Donéstevez,
   Marielis González Muño, and Allan Pierra Fuentes

Development Processes of Open Source Products

Barriers Faced by Newcomers to Open Source Projects: A Systematic
Review 153
*Igor Steinmacher, Marco Aurélio Graciotto Silva, and
Marco Aurélio Gerosa*

Does Contributor Characteristics Influence Future Participation?
A Case Study on Google Chromium Issue Tracking System............ 164
Ayushi Rastogi and Ashish Sureka

A Layered Approach to Managing Risks in OSS Projects.............. 168
*Xavier Franch, Ron Kenett, Fabio Mancinelli, Angelo Susi,
David Ameller, Ron Ben-Jacob, and Alberto Siena*

A Methodology for Managing FOSS Migration Projects............... 172
Angel Goñi, Maheshwar Boodraj, and Yordanis Cabreja

The Agile Management of Development Projects of Software Combining
Scrum, Kanban and Expert Consultation 176
Michel Evaristo Febles Parker and Yusleydi Fernández del Monte

Testing and Assurance on Open Source Projects

An Exploration of Code Quality in FOSS Projects 181
Iftekhar Ahmed, Soroush Ghorashi, and Carlos Jensen

Polytrix: A Pacto-Powered Polyglot Test Matrix 191
Max Lincoln and Fernando Alves

Flow Research SXP Agile Methodology for FOSS Projects 195
*Gladys Marsi Peñalver Romero, Lisandra Isabel Leyva Samada, and
Abel Meneses Abad*

How to Support Newcomers Onboarding to Open Source Software
Projects 199
Igor Steinmacher and Marco Aurélio Gerosa

Global Impact on Open Source Communities and Development

The Census of the Brazilian Open-Source Community 202
Gustavo Pinto and Fernando Kamei

Cuban GNU/Linux Nova Distribution for Server Computers........... 212
*Eugenio Rosales Rosa, Juan Manuel Fuentes Rodríguez,
Abel Alfonso Fírvida Donéstevez, and Dairelys García Rivas*

A Study of the Effect on Business Growth by Utilization and
Contribution of Open Source Software in Japanese IT Companies 216
  Tetsuo Noda and Terutaka Tansho

Case Studies and Demonstrations of Open Source Projects

USB Device Management in GNU/Linux Systems 218
  Edilberto Blez Deroncelé, Allan Pierra Fuentes,
  Dayana Caridad Tejera Hernández, Haniel Cáceres Navarro,
  Abel Alfonso Fírvida Donéstevez, and Michel Evaristo Febles Parker

PROINFODATA: Monitoring a Large Park of Computational
Laboratories... 226
  Cleide L.B. Possamai, Diego Pasqualin, Daniel Weingaertner,
  Eduardo Todt, Marcos A. Castilho, Luis C.E. de Bona, and
  Eduardo Cunha de Almeida

Book Locator: Books Manager 230
  Dairelys García Rivas

Automation of Agricultural Irrigation System with Open Source 232
  Bladimir Jaime Pérez Quezada and Javier Fernández

When Are OSS Developers More Likely to Introduce Vulnerable Code
Changes? A Case Study 234
  Amiangshu Bosu, Jeffrey C. Carver, Munawar Hafiz,
  Patrick Hilley, and Derek Janni

Author Index 237

Code Review Analytics: WebKit as Case Study

Jesús M. González-Barahona[1], Daniel Izquierdo-Cortázar[2], Gregorio Robles[1], and Mario Gallegos[3]

[1] GSyC/LibreSoft, Universidad Rey Juan Carlos
{jgb,grex}@gsyc.urjc.es
[2] Bitergia
dizquierdo@bitergia.com
[3] Universidad Centroamericana José Simón Cañas
mgallegos@uca.edu.sv

Abstract. During the last years, most of the large free / open source software projects have included code review as an usual, or even mandatory practice for changes to their code. In many cases it is implemented as a process in which a developer proposing some change needs to ask for a review by another developer before it can enter the code base. Code reviews, therefore, become a critical process for the project, which could cause delays in contributions being accepted, and risk to become a bottleneck if not enough reviewers are available. In this paper we present a methodology designed to analyze the code review process, to determine its main characteristics and parameters, and to detect potential problems with it. We also present how we have applied this methodology to the WebKit project, learning about the main characteristics of how code review works in their case.

1 Introduction, Motivation and Goals

Code review is gaining importance in free, open source software (FLOSS) projects, as it started to gain relevance several years ago in proprietary software firms [2,1]. Currently, most large FLOSS projects are using it in one way or another. Understanding how it is working, how it can be characterized with traceable, measurable parameters, and understating how it may affect to the relationships between actors in the project is becoming of great importance [4]. In this paper, we present a methodology that addresses these needs[1]. It starts by identifying traces from the review process in software development repositories such as source code management or issue tracking systems, and goes all the way to the characterization of performance and extension properties of the process. In particular, the following research questions are addressed: (Q1) To which extent can the review process based on traces in development repositories be characterized? (Q2) How can the evolution over time of the code review process be characterized?

---

[1] Reproduction information and data sources of the study, according to [3], are available at http://gsyc.es/~jgb/repro/2014-oss-webkit-review

The answer to Q1 is important because if those traces can be found, and automatically extracted from software development repositories, an automated or semiautomatic methodology could be designed, implemented and deployed to track the evolution of the main parameters of the code review process. This would be a first step to build an automated dashboard that allows to better understand the process and for the continuous follow-up of those aspects by any interested party [5]. Q2 is focused on identifying parameters as simple as possible to calculate, but that capture information about important aspects of the evolution of the code review process. Again, if this can be done, instead of using a large collection of complex parameters, a small number automatically computed could be used. In our case, we have checked these two questions in the well-known WebKit project.

The next section presents the WebKit code review process, and provides a qualitative answer to Q1. Then, the methodology for the data retrieval and postprocessing is described in Section 3, including a quantitative answer to Q1. The analysis itself, with the answer to Q2, follows in Section 4. Section 5 is devoted to discussion and the analysis of the main threats to validity. The paper concludes with a section presenting the conclusions.

2 The Code Review Process, and Its Traces

In WebKit, most significant source code contributions must go through a review process. However, activities considered trivial, or very basic maintenance issues (such as minor fixes due peculiarities of one of the platforms) can be committed directly.

2.1 Code Review: Is It Possible to Follow It in Detail?

Anyone may send a contribution to WebKit. But usually the contribution must go through a review process, and be accepted by a reviewer. Both committers and reviewers are selected by previous WebKit committers and reviewers by a meritocratic, peer-approval process[2].

The contribution process[3] is centered around the Subversion repository and the Bugzilla system. Developers start by choosing or opening a new ticket in Bugzilla. The ticket may correspond to a bug report, a feature request, or something else. While working on it, developers compose a patch in their local working copy of the Subversion repository, including entries in changelog files, describing the changes and identifying the ticket. Then, it is submitted to Bugzilla with another script, where it is attached to its ticket.

Usually, upon submission to Bugzilla, code review is requested. This can be done by flagging the attachment to the ticket (as "Review?"), but it can also be requested by other means, such as in the project IRC channel. Unfortunately, only the flagging in Bugzilla leaves traces. Fortunately, flagging the ticket is

---

[2] http://www.webkit.org/coding/commit-review-policy.html
[3] http://www.webkit.org/coding/contributing.html

the most popular requesting method. The review request is not directed to a reviewer in particular, although the developer may try to get the attention of some of them by CCing them in the ticket update, or by directly addressing them somehow.

Reviewers deal with review requests according to their preferences. Once they have reviewed a contribution, they can decide to accept it, to ask for changes, or to reject it. Acceptance and rejection is signaled with a new flag ("Review+" or "Review-") to the ticket. When developers are asked for changes, they have to send a new patch for review with a new "Review?" flag. Changes may follow several iterations, which can in many cases be tracked by examining the review flags.

Once a contribution is accepted, it can be committed to the Subversion repository by any committer, or marked for automatic commit by the commit-queue bot. Therefore, only those developers who are also committers usually commit their own contributions. This means that the "committer" field in the Subversion commit record does not contain information about the real author or reviewer (and Subversion keeps no information about authorship).

2.2 Traces

Summarizing, the traces left by the code review process are:

- Changelog files in the Subversion repository. They include information for every commit, and at least author (usually including name and email address), Bugzilla identifier of the related ticket, and reviewer (if the commit was accepted after a code review process).
- Commit records in the Subversion repository. The committer information is not reliable, but useful information can still be extracted from the committed changelog files.
- Attachments and flags in the Bugzilla repository. Each contribution is usually submitted as an attachment to a ticket, and reviews are requested and granted usually setting flags in it. Detailed timing of all these operations is available, and some information about the person performing it. So, at least this information is available: time of review request ("Review+?") and results of the review process: acceptance ("Review+") or rejection ("Review-"), and Bugzilla identifier of those changing the state of the ticket.

Not always all of this information is available. However, commits with missing information is mainly from old reviews; since about 2005 a very large fraction of them have all data (see details in section 3). Using this information, a characterizations of the review process is possible:

- The most reliable information about authors and reviewers (i.e., name and email address) comes from the changelog files, since they are well documented.
- Alternatively, authors and reviewers could also be determined from their identities in Bugzilla ticketsA manual examination of a random collection shows that both sources offer the same information, as usually scripts automatically include identities in both.

- Timing information is obtained from Bugzilla. The review process starts when a developer flags an attachment to a ticket as "Review?". The end of the process occurs when a "Review+" flag is set. If there are several requests, the timing of each "Review?" flag can also be determined.
- Authors and reviewers are linked to tickets thanks to the ticket identifier found in the changelog files.

With all this information the duration of reviews, and the number of iterations (number of review requests) can be tracked with great accuracy. Therefore, the answer to Q1 is, in principle, positive: the review process can be characterized with great detail at least in the above described terms. Some other aspects of the review process could be characterized with these data, such as the size of reviewed code (which could be extracted both from the commit record and the attachments to the ticket), the changes to code due to the review process (comparison of attachments to the ticket), etc.

3 Methodology

The methodology that we have used to characterize the code review process in WebKit is based on the following steps, similar to those described in [3]: data retrieval from development repositories into databases; clean-up, organization and sampling; and analysis. The first two steps are presented in this section.

3.1 Data Sources and Data Retrieval

The study has been performed using data from the Bugzilla (issue tracking) system[4], and from the Subversion (source code management) repository[5] of the WebKit project. In the case of the Subversion repository, it has been accessed through a git front-end[6] which allows for complete access to all the information.

The git front-end to the Subversion repository was cloned on January 17th 2013 and includes information since August 24th 2001. Metainformation about all commit records (a total of 125,863) was obtained using CVSAnalY, from the MetricsGrimoire toolset[7]. This metainformation was used mainly for cross-validation, but is not really a data source for this study. The git clone was also used to extract information from changelog files. These files, which are spread through the source code tree, were identified, and their relevant information extracted, using an ad-hoc script based in part in the webkitpy library[8], maintained by the WebKit project itself. This script retrieves the complete list of commits from the git clone, identifying and parsing for each of them the modified changelog files. From these files, it extracts the relevant fields: author, reviewer

---

[4] https://bugs.webkit.org
[5] http://www.webkit.org/building/checkout.html
[6] http://trac.webkit.org/wiki/UsingGitWithWebKit
[7] http://metricsgrimoire.github.com
[8] https://trac.webkit.org/browser/trunk/Tools/Scripts/webkitpy

and Bugzilla ticket. A total of 117,079 entries in changelog files were identified this way.

The exact strategy followed by the script to determine changelog entries starts by obtaining a list of all commit identifiers (hashes) directly from the git frontend. For each of them, changelog files added or modified are identified, and their diff information obtained. All entries in those diffs are considered to be related to the commit. The author, reviewer and Bugzilla ticket identifier are retrieved and stored.

Information for all changes to all Bugzilla tickets was retrieved on January 29th 2013 using Bicho, from the MetricsGrimoire toolset. A total of 100,221 issues, and 976,879 ticket state changes were retrieved, with the first ticket dating from June 1st 2005 (there is a single older ticket from 2000, which seems to be an error, and was not considered).

3.2 Cleaning and Organizing the Data

A quick observation of the retrieved dataset shows how, as expected, committer information in the Subversion repository is unreliable. 24,406 commit records were found to have a committer which is not the author, according to the changelog information. Many of those are submitted by bots, who perform automatic commits of already reviewed code, or of small maintenance changes. For example, during 2012 about 22% of commits (7,079 out of 31,923) were performed by bots.

The analysis of the changelog files shows that they covered a very large fraction of all commits: only 8,784, or about 7% of the commits, are missing in the changelogs. The difference can be attributed in some cases to errors, but usually to minor maintenance commits, such as versioning commits, which are not considered to deserve an entry in changelog files. Although other approaches are possible, we considered only a single author for each commit. This meant that of the commits identified in changelog files, an additional 578 (about 0.5%) were ignored because they included information about more than one author (this usually happens when several developers collaborated in the code change).

The number of commits that included "Reviewed by" and similar entries was 74,290. On the other hand, the number of commits with changelog files that reference a Bugzilla ticket is 68,460. Both conditions (being reviewed and referencing a ticket) are fulfilled by only 60,991. Many of the commits that do not comply with both conditions correspond to the period before June 2005, when tickets were not introduced in the Bugzilla database. In some cases, a ticket is referenced in more than one commit: 55,649 unique tickets were found in changelogs. When looking for those tickets, some corresponded to non-public tickets (such as those used in some cases by Apple developers), being 54,501 the total number of tickets that we could use in our study.

The last step in selecting tickets is considering only those with complete information about the review process: we need to know when it was initiated, and when it finished. For that, we selected tickets flagged at least once as "Review?" (review request) and "Review+" (review approved).

Our study will consider tickets with review request after 2006 and before 2013. This will avoid the early days of the project when few information about code review is available in the Bugzilla system, and the final part of the sample, which would distort the final part of the evolution studies. For that period, we have a total of 75,179 commits marked as reviewed in the changelog files. Out of them, 61,867 (82%) reference a Bugzilla ticket, with 56,483 different tickets referenced. Of those, 55,303 (98%) are publicly available. Of those, 53,212 include at least one review request and approval: this is our final sample, which will be used for the following analysis.

4 Analysis Over Time

In order to characterize the evolution of the review process over time, after informal discussion with some WebKit developers, we have used the following parameters as they capture the most relevant aspects of how the code review process is changing:

- Bulk parameters: number of commits subject to code review, number of authors and reviewers involved, per month. They capture the "size" of the review process: how many actors are involved, how many actions ("review processes") they perform.
- Performance parameters: number of iterations (review requests needed) per review, delay from review request to reviewed. They capture how much effort is put into the review process (measured by iterations), and how much delay the process is causing (by measuring time-to-review).

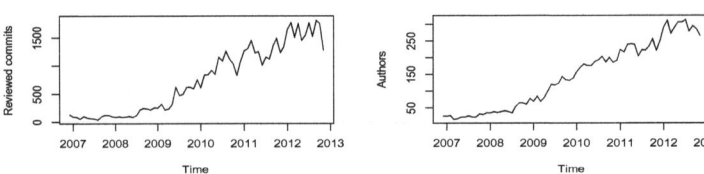

Fig. 1. Bulk parameters (authoring). Total number of tickets corresponding to review processes, and of active authors (review requesters) per month. Time of review request is used to determine the month for each ticket.

Figure 1 shows the bulk parameters for authoring. The number of reviewed commits is increasing clearly over time (from less than 400 per month before mid-2009 to around 1,500 per month during 2012). The number of active authors per month is also increasing, following closely (although not always) the number of reviewed commits. In this case, the growth started a bit earlier than for commits, and shows what at first sight seems to be a linear trend.

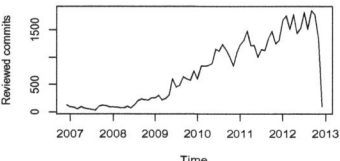
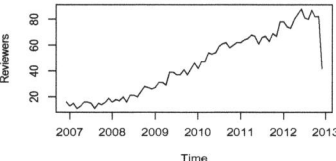

Fig. 2. Bulk parameters (reviewing). Total number of tickets corresponding to review processes, and of active reviewers (review approvers) per month. Time of review approved is used to determine the month for each ticket.

Bulk parameters for reviewing, shown in Figure 2, show very similar patterns. With a median delay of 151 minutes, times for asking for review and granting it are very close, which explains the almost equal shapes for commits. The growth in reviewers, on the contrary, is a bit slower than in the case of authors. While from 2008 to 2012 authors increased from around 50 to 250-300, reviewers grew only from about 20 to 80.

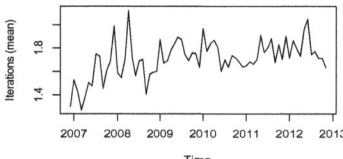
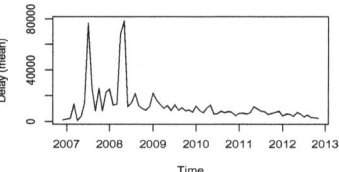

Fig. 3. Performance parameters (mean per month). Number of iterations (review requests for the same ticket) and delay (time from review request to review approval, in minutes).

Performance parameters tell about how the process is actually working over time. Figure 3 shows the evolution of the means: mean iterations and mean delay over time. In the case of iterations (number of cycles implying review requests) per ticket, although there is a lot of variance over time, a slowly growing trend seems clear. In early 2007, the mean number of iterations per ticket was of about 1.4, while in 2012 it remains around 1.8 most of the year.

Despite this increase in the number of iterations, the mean delay for tickets has been decreasing since mid 2008, after a couple of long peaks (in mid 2007 and mid 2008), for which we have found no apparent explanation. Looking at the general trend, it shows how, despite putting more effort in the review process (more iterations), the project is being able of reducing the time-to-review. This means that both reviewers are being more responsive to review requests by authors, but also that those have into account quickly the suggestions for changes, so that a new review cycle can start.

This said, it is important to signal that the distribution of delays is very skewed: while the median for delays is 151 minutes, the mean is 7,447 minutes.

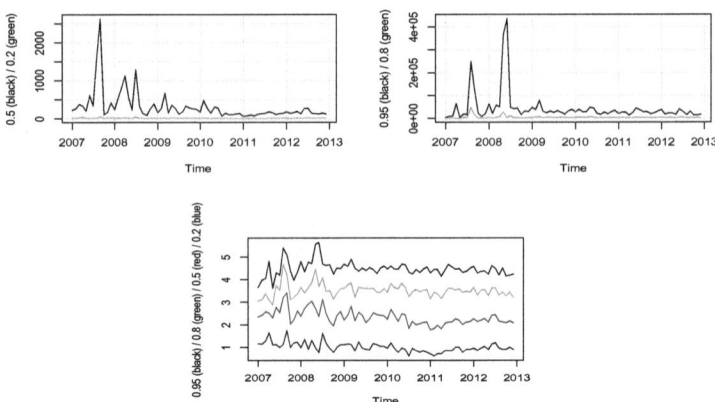

Fig. 4. Performance parameters (quantiles per month). Delay (time from review request to review approval, in minutes) for "quick" (top) and "slow" (middle) reviews, and logarithm of the delay (bottom).

Therefore, we have considered convenient to offer also, in Figure 4, similar information, but now using quantiles. In the top chart in that figure we can see the maximum delay for the quickest closed tickets. For example, the .2 (green) line in that chart shows how the maximum delay for the 20% quickest review processes, over time. The .5 (black) line shows the evolution of the median delay.

For all the quantiles analyzed, the evolution of delay over time is quite similar, and consistent with the one found for the mean delay. Maybe the quickest tickets are reducing their delays, while the slower ones tend to be more stable. This can be seen in the middle chart, with the delays for 80% and 95% of the tickets, but more clearly in the bottom one, which shows the logarithm of delay over time: the red and blue lines shows a tendency to descend since 2008, while the black and green ones are almost horizontal. The bottom chart, taking into account the log scale, shows also the great skewness of the distribution of delays.

5 Discussion and Threats to Validity

In large projects where many different actors contribute, each with their own and usually competing interests, many software development processes are difficult, yet important to understand. In the case of our study, the code review process is specially important, because it controls what enters the code base, but also what is left out. It is a barrier that developers have to overcome for contributing. Therefore, understanding it is really important, and more when the stakeholders are companies competing in the marketplace, but collaborating in the project.

A pure qualitative understanding, based on the modeling of the mechanics of the process, is not enough. Quantitative information is needed to back discussions with data, to detect early problems, and to be able of evaluating solutions and new policies. In this respect, we have quantified several aspects of the process, and have validated them with WebKit developers.

The main contribution of our study, from this point of view, is a detailed methodology, that can be used in the WebKit project and, with some variations, in other projects too. The parameters and charts presented can be the basis for a specialized dashboard that tracks the details of how the review process is evolving. The parameters presented, and the way they are calculated, can also be the basis of a validation system for any quantitative model of the code review process.

There are several threats to the internal validity of the study. The main one is probably the validity of the parameters selected to characterize the code review process. In general, both practitioners and academics consulted agree on the validity and usefulness of them as they can be linked to important concepts such as effort or delay in actions. But more research is needed to really correlate them with other parameters, so that it becomes clear that they are really important for the review process. Other threats to internal validity are related to the actual data retrieval process, the validity of the analyzed sample, and the exact procedures for estimating the parameters. In general, all of them have been validated with developers from the project. Section 3, and the answer to Q1, have also tried to establish how the sample is good and large enough, as well as the process for estimating parameters from it. However, errors and conceptual problems may remain.

With respect to external validity, it is important to notice that this study does not try to state proprieties to be valid in other projects, nor even in the future of WebKit. We have only tried to determine techniques and artifacts that help to understand the review process, and not to determine general laws or models of how it works. This said, the methodology and the presented artifacts (charts, statistics) are meant to be valid for other projects, and therefore threats to external validity can be applied to them.

6 Conclusions

This paper has presented a detailed methodology for the quantitative analysis of the code review process in large software development projects, based on traces left in software repositories. The methodology has been tested with WebKit, a large and complex project with high corporate involvement. Some developers have given us assistance in the validation and understanding of the code review process, ensuring a higher usability of the results for its stakeholders.

We have answered the two research questions stated in the introduction of this paper. First of all, we have determined how there is enough information in the project repositories to characterize the code review process, and how it can be used, in fact, to calculate parameters that seem to be related to the extension and performance of the project (Q1). We have also characterized the evolution over time of the process using quantitative (bulk or performance) parameters, and have shown how they can be useful to understand such evolution (Q2).

By doing this we have characterized the review process of the WebKit project, and proposed the fundamentals for a dashboard that can serve to evaluate it. We have found that the importance and extension of code review is growing in the project, that reviewers are not growing as fast as authors - and although this has not supposed delays so far, it could cause bottlenecks in the future. On the contrary, the project is improving in the code review process over time, probably due to developers devoting more effort to it.

Acknowledgments. The work of Gonzalez, Robles and Izquierdo has been funded in part by the Spanish Gov. under SobreSale (TIN2011-28110) and Torres Quevedo (PTQ-12-05577). We thank the Webkit developers for their feedback and suggestions.

References

1. Ackerman, A.F., Buchwald, L.S., Lewski, F.H.: Software inspections: An effective verification process. IEEE Software 6(3), 31–36 (1989)
2. Ackerman, A.F., Fowler, P.J., Ebenau, R.G.: Software inspections and the industrial production of software. In: Proc. of a Symposium on Software Validation: inspection-Testing-Verification-Alternatives, pp. 13–40. Elsevier Inc. (1984)
3. Jesus, M.: Gonzalez-Barahona and Gregorio Robles. On the reproducibility of empirical software engineering studies based on data retrieved from development repositories. Empirical Software Engineering 17, 75–89 (2012)
4. Rigby, P.C., German, D.M., Storey, M.A.: Open source software peer review practices: a case study of the apache server. In: Proceedings of the 30th International Conference on Software Engineering, pp. 541–550. ACM (2008)
5. Rigby, P.C., Storey, M.A.: Understanding broadcast based peer review on open source software projects. In: Proceedings of the 33rd International Conference on Software Engineering, pp. 541–550. ACM (2011)

Navigation Support in Evolving Open-Source Communities by a Web-Based Dashboard

Anna Hannemann, Kristjan Liiva, and Ralf Klamma

RWTH Aachen University,
Ahornstrasse 55, 52056 Aachen, Germany
{hannemann,liiva,klamma}@dbis.rwth-aachen.de

The co-evolution of communities and systems in open-source software (OSS) projects is an established research topic. There are plenty of different studies of OSS community and system evolution available. However, most of the existing OSS project visualization tools provide source code oriented metrics with little support for communities. At the same time, self-reflection helps OSS community members to understand what is happening within their community. Considering missing community-centered OSS visualizations, we investigated the following research question: Are the OSS communities interested in a visualization platform, which reflects community evolution? If so, what aspects should it reflect?

To answer this research question, we first conducted an online survey within different successful OSS communities. The results of our evaluation showed that there is a great interest in community-centered statistics. Therefore, we developed an OSS navigator: a Web-based dashboard for community-oriented reflection of OSS projects. The navigator was filled with data from communication and development repositories of three large bioinformatics OSS projects. The members of these OSS communities tested the prototype. The bioinformatics OSS developers acknowledged the uniqueness of statistics that the NOSE dashboard offers. Especially, graph visualization of the project social network received the highest attention. This network view combined with other community-oriented metrics can significantly enhance the existing visualizations or even be provided as a standalone tool.

1 Introduction

Success of an OSS project is tightly interwoven with the success of its community [Ray99], [HK03]. OSS systems co-evolve strongly with their communities [YNYK04]. Thus, the more successful a project is, the higher is the degree of its complexity in terms of project structure and community size. The complexity affects the awareness of community members of what is happening in their community. In interviews with OSS developers Gutwin et al. in [GPS04] find out that the awareness of other developers within OSS projects is essential for an intact project life. Within the study, project mailing lists (MLs) and text chats are determined as the main resources for maintaining group awareness. However, in large OSS projects, it gets very difficult for community members, especially for the less experienced ones, to establish a complete and correct perceptional awareness model. In such cases, Gutwin et al. suggest to develop new representation methods for communication and its history.

Considering OSS mining research, there are already studies concentrating on OSS communication analysis: to investigate social network structure [BGD+06], to analyze content [BFHM11], to estimate the sentiment within OSS communities [JKK11]. OSS communication repositories reflect complete communities of the corresponding projects. In contrast, OSS source code repositories are restricted to the developers only. If we take a look at the OSS visualization platforms (e.g. GitHub, Ohloh, etc.), then they are focused either on source code or individual contributors. Platforms which provide OSS metrics based on project communication are still missing. To investigate this research niche, we address the following research question: **Are the OSS communities interested in a platform reflecting community evolution and if so, what evolution aspects should it reflect?**

The rest of the paper is organized as follows. Section 2 provides an overview on related systems for OSS project evolution visualization. To address our research questions, we executed an iterative study (Section 3). The achieved results are presented in Section 4. Section 5 concludes the paper and gives an overview of some ideas for future work.

2 Related Research

There are already plenty of related applications available for OSS development visualization. To give an overview of existing concepts and principles, the more notable ones are presented.

GitHub[1] offers a web-based hosting for software projects. Additionally, it provides visualizations focused mainly on project source code (commit activity, code amount) and some statistics on project contributors (contributor activity, followers and following people, projects, organizations, etc). Another popular web-platform for software projects' hosting is **SourceForge**[2]. It offers just some statistics on project traffic (hits on the project, number of downloads) and SVN activity. What statistics are visible to users depends on the project settings. In contrast, a web-service **Ohloh**[3] does not host the actual source code, but simply crawls and analyzes the OSS data. Ohloh offers many charts regarding the source code and contributors (their ranking and activity). Pure statistics are transformed into textual statements. Ohloh also provides data on project estimation effort based on the COCOMO model for software cost development [Boe81]. The next web front-end **Melquiades**[4] provides visualization for the data collected within **FLOSSmetrics**[5] research project [HIR+09]. The supported analysis is

---

[1] GitHub, https://github.com, last checked 2013/09/10
[2] SourceForge, https://sourceforge.net, last checked 2013/09/10
[3] Ohloh, www.ohloh.net, last checked 2013/09/10
[4] Melquiades, melquiades.flossmetrics.org, last checked 2013/09/10
[5] FLOSSmetrics, flossmetrics.org, last checked 2013/09/10

divided into three different types according to data resource used: data from source code repositories, data from mailing lists archives and data from tracker system repositories. However, not all projects have data regarding all three resources. Melquiades offers important metrics, like activity over time, growth and member inflow rate. The next two visualizations **Open Source Report Card (OSRC)**[6] and **Sargas** [SBCS09] provide contributor-oriented metrics. Based on the data from GitHub OSRC establishes developers profiles based on their daily and weekly activities, project participation, etc. Whereas, Sargas estimates the social profile of contributors based on their behavior within four social networks: open discussion forum, developers discussion list, discussions about the bugs and social network extracted from the source code.

To summarize, the existing applications are ranging from source code hosting services with visualization tools to pure analysis and visualization platforms. The last clearly proves the need and the interest of the OSS communities in self-monitoring tools. However, the existing systems focus mainly either on the system source code or on individual contributors. For monitoring of community evolution the information need to be presented from different perspectives.

3 Study Settings

Figure 1 represents the workflow of our study. To find out if the OSS members are interested in the community-related reflection of their projects, we first conducted an online survey within OpenStack, PostgreSQL, GIMP, Mozilla, Oracle VM VirtualBox, GNOME, TomCat OSS communities. The survey addressed questions related to the developer interest in a community-oriented metrics and what metrics are missing in the existing OSS navigators. Most of the questions had an optional comment field. We contacted OSS developers via the Internet Relay Chat (IRC) channels. The survey was anonymous, therefore, it was not possible to trace from which project the participants originated. Nevertheless, based on the survey results and by observing each chat for the next four hours, no malicious users were detected. The result of the survey was a positive answer to the first part of our research question. The OSS members do have strong interest in platforms reflecting OSS community evolution. Additionally, the OSS members suggested several ideas for metrics/aspects, which were assumed to be important to be aware of. To evaluate the feasibility of the collected ideas, we next applied prototype testing.

We selected a dashboard as a technological approach. *"Dashboard is a visual display of the most important information needed to achieve one or more objectives; consolidated and arranged on a single screen so the information can be monitored at a glance"* [Few06]. The goal of the developed Navigator for OSS Evolution (NOSE) is to provide community-oriented navigation support for OSS project members. We filled the NOSE dashboard with the data from three long-term bioinformatics OSS (BioJava, Biopython and BioPerl), which have been already analyzed in our previous studies (e.g. [HK13]). Therefore, we

---

[6] Open Source Report Card, osrc.dfm.io, last checked 2013/09/10

were able to proceed with the prototype testing immediately after the development. An iterative development process of the NOSE dashboard was executed. Before starting the survey with bioinformatics communities, the dashboard was evaluated with 10 computer scientists. This evaluation was used to identify the design shortcomings of the developed dashboard.

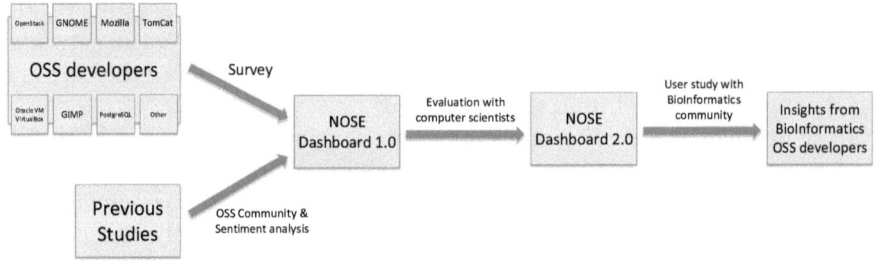

Fig. 1. Study Workflow

The relevance of the presented metrics and the data quality reflected in NOSE were directly investigated within bioinformatics OSS communities. We contacted the bioinformatics OSS community members via private emails. Such textual inquires encourage the participants to explain their answers and, thus, provides a more detailed feedback. Moreover, informal email exchange could trigger a fruitful discussion. The sent out email consisted of a short description of the NOSE goal and three questions:

– Do you find that the visualization features offered by GitHub are sufficiently informative?
– Would you additionally like to have features to represent the community and its structures?
– If so, would network graph be a viable option?

4 Navigator for OSS Community: Evaluation Results

In following the results of both conducted surveys and prototype testing are presented.

4.1 Survey of OSS Developers

From OSS developers we received 32 responses with many (49) comments. Some developers provided initial feedback directly in the chat. Everyone who started the survey also finished it, what indicates the true interest of the participants in the survey. Figure 2 displays the survey questions with the collected feedback. Every question was optional, therefore some questions had less than 32 answers.

Navigation Support in Evolving Open-Source Communities   15

Web-platforms like Ohloh or GitHub were used by 75% of the participants. There were 16 comments in total, with GitHub being mentioned 13 times, SourceForge 5 times and Ohloh 4 times. 63.3% of the OSS developers were interested in the statistics related to community evolution. However, in four of the seven comments the participants mentioned, that it was unclear what kind of information was meant or *"How would/could I benefit from those information?"*

Social Analysis. The OSS developers were mostly interested in getting statistics from the MLs regarding the whole community. MLs were recognized as a useful source of information for getting an approximate user base size. However, one participant also mentioned a negative aspect, that *"[...] too much statistical evaluation could put the community of as they feel 'observed' "*. The most opinions of

Question	Yes	No	Optional
Do you use (F)OSS?	32 (100%)	0 (0%)	
Have you ever contributed to any (F)OSS project (e.g bug report)?	31 (96.9%)	1 (3.1%)	
Ohloh, GitHub and similar platforms focus on LoC/commit statistics, whereas I want to focus on community(=people) evolution. Do you find the information to be interesting?	19 (63.3%)	11 (36.7%)	
I want to provide statistics on the whole community, but also every contributor from the mailing list. Would you find it useful?	24 (75%)	8 (25%)	
Are you interested in a network view (social graph) of your (F)OSS community?	23 (71.9%)	9 (28.1%)	
Are you interested in a text mining analysis of your (F)OSS project communication (e.g. determine main topics mailing list)	20 (64.5%)	11 (35.5%)	
Are you interested in sentiment analysis, estimating the "mood" of each message and providing aggregated statistics for the whole project?	10 (31.3%)	22 (68.8%)	
Do you use web-services like Ohloh, GitHub or other similar platforms?	24 (75%)	5 (15.6%)	Not currently, but I have done so in the past: 3 (9.4%)
Would a personalized dashboard (e.g. choosing data, visualization form etc.) be more useful than the project-oriented view offered by Ohloh and GitHub?	8 (25%)	2 (6.3%)	Need to try it first: 22 (68.8%)
Is there any statistics regarding your (F)OSS project that you would like to have? For example, if you have used GitHub or Ohloh and it was missing something important.			Free form answers

Fig. 2. OSS Developers Survey Results

network graph representation of an OSS community were again positive with only two answerers mentioned that they would find it more interesting than useful.

Text Mining (TM). Communication presents not only a source for social network analysis. It also provides a great unplugged pool for TM. TM methods allow to determine end-user requirements, discover conflicts, etc. Special area of TM is sentiment analysis. The mood of a user can be implicitly estimated based on opinionated documents generated by the user (e.g. postings in MLs). However, the OSS developers were mostly uninterested in sentiment analysis. The negative reaction could be the result of little awareness of the sentiment analysis meaning. For example, one of the participants said that *"Only slightly interested. I doubt that much useful information could be drawn from such an analysis, but I would need to know more about the methodology and findings to be sure, and it sounds interesting at least"*. Another participant expressed the concern that such analysis *"Would be interesting/fun, but i m not sure whether its useful for me [...]"*. However, there were also participants that were clearly in favor of sentiment analysis: *"Definitely. I would have stopped before entering some projects if I would have known about the mood swings of their contributors beforehand..."*. In contrast, other responses were clearly against it, for example *"I don't think public statistics of the form "messages from developer X are mostly aggressive" would do anything good"*. The concern of feeling observed was already mentioned in the context of social analysis. Consequently, the developers are rather interested in the aggregated statistics. For example: *"I think this [sentiment analysis] is only useful if combined with a certain segmentation of the user groups"*.

Majority of the participants were interested in TM analysis of the ML communication for other purposes:

- *"[...] determine the needs of the users in addition to voting and tagging in bugtrackers"*
- *"[...] creating FAQs for new contributors"*
- *"[...] finding out in which direction the community wants to evolve"*

Missing Functionalities. Finally, the following statistics were missed in the existing OSS visualizations:

- *"[...] which wiki pages are consulted often, which problem appear in lists/ forums frequently and so on."*
- *"Some projects [...] allow Users to be credited in commits for there Testing or Bug reports (e.g. Reported-By: Tested-By:) including those contributions in statistics would by nice [...]"*
- *"Last time on ohloh I wanted to see a simple list of contributions, but I only found timelines and such, which I find hard to browse"*
- *"More informations about project activity. Most FOSS project have stalled development, are abandoned. This is for me the #1 FOSS problem"*

Dashboard. A personalized dashboard was considered useful by 25% of the participants, while the majority of the participants (68.8%) replied that they would need to try it first. One participant mentioned that *"The projects I contribute to have their own dashboards, I don't think an external dashboard would match my expectations"*. Indeed, many of the OSS hosting platforms offer their own built-in analysis and visualization.

Summarized, the OSS developers showed a strong interest in both social and text-based community communication analysis. To find out which statistical charts and designs were truly useful, there had to be an application that the OSS developers could try out.

4.2 Bioinformatics OSS Developers

Figure 3 displays a screenshot of the NOSE dashboard evaluated by bioinformatics OSS developers. It consists of five widgets: inflow vs. outflow of members in project MLs, number of commits, sentiment within community vs. commit activity, size of community core, social network graph of the project community. The last widget additionally provides several options: to search for a person, to select a yea or a release, and to highlight the core.

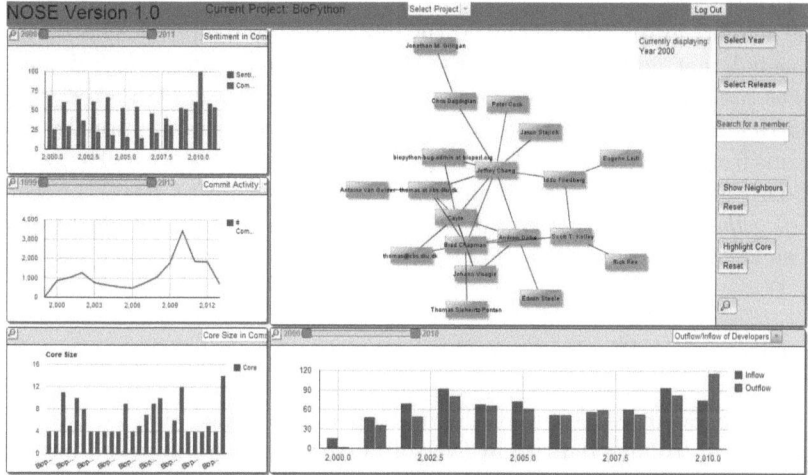

Fig. 3. Screenshot of the NOSE Dashboard Prototype

The survey was sent to members of all three bioinformatics communities. We selected the participants who were active in the MLs in the last two years. In total, we sent an email to 46 project participants. From the 46 persons, nine replied to our email. The participants were also open for the discussion, thus, many issues got an extensive clarification. One of the BioPython developers even posted an article about the NOSE dashboard to his blog[7], which again supports the community interest in the self-monitoring and -reflection.

The proposed community metrics mainly received a positive feedback from the bioinformatics OSS developers. However, three bioinformatics developers reported that, they were not using the OSS visualizations. Nevertheless, one of them mentioned that the NOSE dashboard looked fairly attractive, but he was unsure what he would use it for. Another developer gave a longer explanation saying that *"I don't find this type of community information particularly useful. Over time I've developed a habit of doing my own, informal, reputation scores*

---

[7] 'The Bio* projects: a history in graphs",
http://bytesizebio.net/2013/09/07/bio-projects-a-history-in-graphs,
last checked 2013/11/27

in my head, based on people's list participation and tone, their code, their constructive criticisms, etc.... Who's talking to whom, etc... has never been particularly useful". Similar skepticism among OSS developers was previously reported in [GPS04]. Despite some critical opinions, other evaluation participants were more in favor of the community statistics. One BioJava developer stated, that although BioJava project currently uses Ohloh for visualization, the NOSE dashboard could be complementary to existing visualization platforms. The developer added that these kinds of statistics would be useful for recruiting and funding. *"For instance, we use these types of stats when applying for Google Summer of Code sponsorship"*.

The network graph received by far the most praise and interest. One developer commented that adding *"[...] the social graph to the existing GitHub facilities would be valuable"*. Another reply was: *"The social aspects of OSS projects are no less intriguing than the technological ones!"*. The developer additionally named two gains from such statistics. Firstly, that visualization platforms like the NOSE dashboard are great for getting an overview of the project's history. Secondly, that *"[...] there is a lot to learn from this on how OSS projects get off the ground, what makes a successful project, etc"*.

4.3 Discovered Weaknesses

More Data Sources. Many of the developers mentioned that additionally it would be nice to have data from GitHub. One developer expressed interest in comparing the social networks created based on GitHub and ML. GitHub is *"a great place to discuss code specifics, so is often easier to go back and forth on than writing e-mails. It would be cool to see how the interactions there overlay on this"*. Another developer expressed interest in getting such communication statistics from the project LinkedIn[8] group.

Network Graph. Broadcasts were excluded from the network graph visualization. If the broadcasts were included, it would create an enormous amount of edges. That would make the rendering of a comprehensible network almost impossible. Additionally, it would create many hubs, thus lowering the presence of actual core developers. Further, one of the evaluation participants identified one core developer, who was split into two aliases. This splitting decreased his/her social role in the network graph. Nevertheless, bioinformatics developers believed the network graph captures the community quite accurately.

There were many suggestions and requests. Most of the feature requests were directed at getting statistics from additional sources and not only from the ML. Some metrics requests were related to the social network graph:

- *"[...] add a graph to the dashboard that shows, for each year who are the top linked nodes"*

---

[8] http://www.linkedin.com/groups/BioJava-58404?home=&gid=58404&trk=anet_ug_hm, last checked 12.09.2013

- *"[...] graph comparative metrics, such as consensus linkage, slope of the edge-number histogram [...]"*. The developer suggested that it could be used for comparing the three bioinformatics projects.
- *"[...] use the graph's connecting edges to indicate the strength/weight of the the connection (i.e. line thickness linked to number of email conversations)"*

5 Conclusions and Future Work

In this paper, we addressed the research question: Are the OSS communities interested in a visualization platform, which reflects community evolution? If so, what aspects should it reflect? To answer this research question, we surveyed members from different OSS communities. Based on the survey results, we developed a dashboard prototype for community-oriented navigation in OSS projects. The evaluation within three long-term bioinformatics OSS showed a strong interest of OSS developers in visualization of community statistics. Especially, the network graph visualization of the communities was recognized as the most interesting metric. The developers are more interested in aggregated statistics in order to avoid the feeling of being observed among the project participants. On contrary, sentiment analysis did not get much attention, which might be a result of a poor description or little awareness of the analysis method. However, some evaluation participants saw the NOSE platform more as fun, than as a useful evolution barometer. Further, the dashboard was suggested as a possible extension for the existing platforms and not as a standalone application.

Our next steps are to realize the identified requirements. In terms of analysis metrics, the OSS members wish topic-based text mining [GDKJ13] measures, with the goal to see where users struggle. Considering the data, there are many requests to extend the data sources, for example by the data from GitHub. Further studies with domains outside bioinformatics are needed to achieve truly generalizable results. Currently, we apply the concept of the NOSE dashboard to support and manage an OSS community around a EU project Learning Layers[9].

Acknowledgement. This work is supported by the LAYERS FP7 ICT Integrated Project (grant agreement no. 318209) of the European Commission.

References

[BFHM11] Bohn, A., Feinerer, I., Hornik, K., Mair, P.: Content-based social network analysis of mailing lists. The R Journal 3(1), 11–18 (2011)
[BGD+06] Bird, C., Gourley, A., Devanbu, P., Gertz, M., Swaminathan, A.: Mining email social networks. In: Proceedings of the 2006 International Workshop on Mining Software Repositories, MSR 2006, pp. 137–143. ACM, New York (2006)
[Boe81]   Boehm, B.W.: An experiment in small-scale application software engineering. IEEE Transactions on Software Engineering 7(5), 482–493 (1981)

---

[9] Learning Layers, http://learning-layers.eu, last checked 2013/11/10

[Few06] Few, S.: Information Dashboard Design: The Effective Visual Communication of Data, p. 35. O'Reilly Media (2006)

[GDKJ13] Günnemann, N., Derntl, M., Klamma, R., Jarke, M.: An interactive system for visual analytics of dynamic topic models. Datenbank-Spektrum 13(3), 213–223 (2013)

[GPS04] Gutwin, C., Penner, R., Schneider, K.: Group awareness in distributed software development. In: Proceedings of the 2004 ACM Conference on Computer Supported Cooperative Work, CSCW 2004, pp. 72–81. ACM, New York (2004)

[HIR+09] Herraiz, I., Izquierdo-Cortazar, D., Rivas-Hernandez, F., Gonzalez-Barahona, J., Robles, G., Duenas-Dominguez, S., Garcia-Campos, C., Gato, J.F., Tovar, L.: Flossmetrics: Free/libre/open source software metrics. In: 13th European Conference on Software Maintenance and Reengineering, CSMR 2009, pp. 281–284 (2009)

[HK03] von Hippel, E., von Krogh, G.: Open source software and the "private-collective" innovation model: Issues for organization science. Journal on Organization Science 14(2), 208–223 (2003)

[HK13] Hannemann, A., Klamma, R.: Community dynamics in open source software projects: Aging and social reshaping. In: Petrinja, E., Succi, G., El Ioini, N., Sillitti, A. (eds.) OSS 2013. IFIP AICT, vol. 404, pp. 80–96. Springer, Heidelberg (2013)

[JKK11] Jensen, C., King, S., Kuechler, V.: Joining free/open source software communities: An analysis of newbies' first interactions on project mailing lists. In: Proceedings of the 44th Hawaii International Conference on System Sciences (HICSS), pp. 1–10 (January 2011)

[Ray99] Raymond, E.S.: The Cathedral and the Bazaar. O'Reilly Media (1999)

[SBCS09] de Sousa, S.F., Balieiro, M.A., dos R. Costa, J.M., de Souza, C.R.B.: Multiple social networks analysis of floss projects using sargas. In: 42nd Hawaii International Conference on System Sciences, HICSS 2009, pp. 1–10 (2009)

[YNYK04] Ye, Y., Nakakoji, K., Yamamoto, Y., Kishida, K.: The co-evolution of systems and communities in free and open source software development. In: Koch, S. (ed.) Free/Open Source Software Development, pp. 59–82. Idea Group Publishing, Hershey (2004)

Who Contributes to What? Exploring Hidden Relationships between FLOSS Projects

M.M. Mahbubul Syeed[1] and Imed Hammouda[2]

[1] Department of Pervasive Computing,
Tampere University of Technology, Finland
mm.syeed@tut.fi

[2] Department of Computer Science and Engineering,
Chalmers and University of Gothenburg, Sweden
imed.hammouda@cse.gu.se

Abstract. In this paper we address the challenge of tracking resembling open source projects by exploiting the information of which developers contribute to which projects. To do this, we have performed a social network study to analyze data collected from the Ohloh repository. Our findings suggest that the more shared contributors two projects have, the more likely they resemble with respect to properties such as project application domain, programming language used and project size.

1 Introduction

With the exponential increase of Free/Libre Open Source (FLOSS) projects [6], searching for resembling open source components has become a real challenge for adopters [7]. By **resemblance**, we mean similarity factors between projects such as features offered, technology used, license scheme adopted, or simply being of comparable quality and size levels.

In this paper we exploit the information of 'which developers contribute to which FLOSS projects' to identify resembling projects. Our assumption is that if a developer contributes to several projects, simultaneously or at different times, then there might be implicit relationships between such projects. For example, one FLOSS project may use another as part of its solution [17] or two projects could be forks of a common base [16].

The idea of collecting and studying data of who contributes to which FLOSS projects is not new. The question has been the focus of many studies due to its relevance from many perspectives. For instance, the question has been significant for companies who want to identify who influences and controls the evolution of a specific project of interest[2], or to explore the social structure of FLOSS development [1], or simply to study what motivates people to join open source communities [3]. In this work, we address the following research questions:

1. How do FLOSS project development communities overlap?
2. To what extent can developer sharing in FLOSS projects approximate resemblance between the projects themselves?

For answering these questions, we used social network analysis techniques to analyze data collected from the Ohloh repository [4].

2 Study Design

This section presents in detail our study design, covering discussion on the data sources, required data sets, data acquisition, cleaning, and analysis process along with validation and verification of the analysis process.

2.1 Data Source

For this study we selected Ohloh data repository [4], which is a free, public directory of open source software projects and the respective contributors.

Ohloh collects and maintains development information of over 400 thousand FLOSS projects, and provide analysis of both the codes history and ongoing updates, and attributing those to specific contributors. It can also generate reports on the composition and activity of project code bases. These data can be accessed and downloaded through a set of open API which handles URL requests and responses [4]. The response data is expressed as an XML file, an example of which is shown in Fig. 1.

Our selection of Ohloh data repository is predominantly influenced by the following factors: (a) Ohloh data can be publicly reviewed, which in turn makes it one of the largest, most accurate, and up-to-date FLOSS software directories available; (b) the use of Ohloh repository makes FLOSS data available in a cleaned, unified and standard platform independent format. This makes the process of data analysis and visualization independent of technology, and data repository.

2.2 Data Collection

The following information has been collected from Ohloh repository in relation to this study:

Developer Account Information: An Account represents an Ohloh member, who is ideally a contributor to one or more FLOSS projects. Ohloh records a number of properties (or attributes) for an account. Among thousands of registered members, we collected account information of top 530 contributors according to assigned kudo rank of 10 and 9. Kudo rank is a way of appreciation to the FLOSS contributors through assigning a number between 1 and 10 for every Ohloh account [5].

Project Data: A Project represents a collection of source code, documentation, and web site data presented under a set of attributes, a list of which can

```
- <project>
   <id>1</id>
   <name>Apache Subversion</name>
   <url>http://www.ohloh.net/p/subversion.xml</url>
   <html_url>http://www.ohloh.net/p/subversion</html_url>
   <created_at>2006-10-10T15:51:31Z</created_at>
   <updated_at>2013-08-14T08:04:55Z</updated_at>
 + <description></description>
   <homepage_url>http://subversion.apache.org/</homepage_url>
   <download_url>http://subversion.apache.org/packages.html</download_url>
   <url_name>subversion</url_name>
   <medium_logo_url>http://cloud.ohloh.net/attachments/1/svn_med.png</medium_logo_url>
   <small_logo_url>http://cloud.ohloh.net/attachments/1/svn_small.png</small_logo_url>
   <user_count>8463</user_count>
   <average_rating>4.22561</average_rating>
   <rating_count>2265</rating_count>
   <review_count>16</review_count>
   <analysis_id>15032654</analysis_id>
 + <tags></tags>
 - <analysis>
     <id>15032654</id>
     <url>http://www.ohloh.net/analyses/15032654.xml</url>
     <project_id>1</project_id>
     <updated_at>2013-08-14T00:03:27Z</updated_at>
     <logged_at>2013-08-14T00:00:13Z</logged_at>
     <min_month>2000-03-01T00:00:00Z</min_month>
     <max_month>2013-08-01T00:00:00Z</max_month>
     <twelve_month_contributor_count>37</twelve_month_contributor_count>
     <total_code_lines>562264</total_code_lines>
   + <factoids></factoids>
   + <languages graph_url="http://www.ohloh.net/p/subversion/analyses/15032654/languages.png">
     <main_language_id>42</main_language_id>
     <main_language_name>C</main_language_name>
   </analysis>
 + <licenses></licenses>
```


(a)


```
- <position>
   <title>lead developer</title>
   <organization/>
 + <html_url></html_url>
   <created_at>2010-03-20T21:17:45Z</created_at>
   <started_at/>
   <ended_at/>
 + <sparkline_url></sparkline_url>
   <commits>11747</commits>
 - <project>
     <id>3180</id>
     <name>TortoiseSVN</name>
     <url>https://www.ohloh.net/p/tortoisesvn.xml</url>
     <html_url>https://www.ohloh.net/p/tortoisesvn</html_url>
     <created_at>2006-10-17T06:39:39Z</created_at>
     <updated_at>2013-08-09T22:14:44Z</updated_at>
   + <description></description>
     <homepage_url>http://tortoisesvn.net</homepage_url>
     <download_url>http://tortoisesvn.net/downloads</download_url>
     <url_name>tortoisesvn</url_name>
   + <medium_logo_url></medium_logo_url>
   + <small_logo_url></small_logo_url>
     <user_count>3567</user_count>
     <average_rating>4.53501</average_rating>
     <rating_count>957</rating_count>
     <review_count>5</review_count>
     <analysis_id>14884957</analysis_id>
   + <tags></tags>
   - <licenses>
     - <license>
         <name>gpl</name>
         <nice_name>GNU General Public License v2.0 or later</nice_name>
       </license>
     </licenses>
   </project>
  </position>
```


(b)

Fig. 1. (a) Project Information (b) Developer Position Information

be found in Fig. 1(a). We collected information of 4261 projects to which a total of 530 developers have contributed.

Position Information: A position is associated with each Ohloh account, which represents the contributions that the account holder has made to the project(s) within Ohloh. Information maintained in a position repository can be found in Fig. 1(b). We collected the position information for each of the 530 contributors.

For downloading these repository data, we have implemented Java programs, one for each repository. Each Java program implements the API corresponding to a repository by combining the repository URL's and the unique API key to query the database. The result data set is in XML format.

2.3 Data Processing

From the collected repository data (i.e., the XML files as presented in Section 2.2), we parsed only the information that has significance to this study. The parsed information is recorded under a defined set of attributes/tags in XLSX files, one for each XML repository file.

Collected information is then merged to built a complete database required for data analysis. A partial snapshot of this database can be visualized in Fig. 2. Each row presents detailed information about (a) a contributor, (b) a project in which he/she contributed, and (c) the record of contribution to that project.

To automate data parsing and merging, parsers and data processors were written in Java. These programs use Jsoup HTML parser [9] and Apache POI [8] for parsing the XML files and to create database in XLSX format, respectively.

id	name	kudo_rank	position	TotalProject Contributed	totalCommits	projectID	projectName	projectUser Account	projectAvgRating	projectLisence	tweveMonthContributionByAll	totalLineOfCode	mainLanguage
337	Stefan Küng	10	2	18	11747	3180	TortoiseSVN	3567	4.53501	gpl	11	221157	C++
337	Stefan Küng	10	2	18	752	9739	CommitMonitor	139	4.33333	gpl	2	10569	C++
337	Stefan Küng	10	2	18	178	616468	CryptSync	14	5.0	gpl	3	3411	C++
337	Stefan Küng	10	2	18	117	487231	EvlmSync	1	5.0	gpl3	0	46162	C#
60504	XhmikosR	9	103	26	1	4482	FFmpeg	894	4.36364	lgpl21	246	699386	C
60504	XhmikosR	9	103	26	402	12790	StExBar	51	4.66667	gpl	3	13952	C++
60504	XhmikosR	9	103	26	27	616468	CryptSync	14	5.0	gpl	3	3411	C++
60504	XhmikosR	9	103	26	109	9739	CommitMonitor	139	4.33333	gpl	2	10569	C++
60504	XhmikosR	9	103	26	339	3180	TortoiseSVN	3567	4.53501	gpl	11	221157	C++

Fig. 2. Partial Snapshot of the database

2.4 Data Analysis

Data analysis targeting to answer the research questions composed of two steps:

First, we created an **Implicit Network** in which two projects have a relationship if both are contributed to by the same contributor. An edge weight in this network represents the number of such common contributors between two projects. As an illustration, consider contributors *Stefan Küng* and *XhmikosR* in Fig. 2. The former contributor has contributed to 4 FLOSS projects as listed under the *projectName* column, while the latter has contributed to 5 projects. Both developers contributed to projects *TortoiseSVN*, *CryptSync*, and *CommitMonitor*. In total, Fig. 2 lists 6 distinct projects. An implicit network among these 6 projects is shown in Fig. 3. Projects *TortoiseSVN*, *CryptSync*, and *CommitMonitor* are linked with edge of weight 2 (i.e. 2 shared developers). All other edges have weight 1 as those project pairs have only one shared contributor. For example, *FFmpeg* and *CommitMonitor* have only developer *XhmikosR* in common.

The complete network is composed of 194424 edges between 4261 projects, with edge weight varies between 1 and 41. Complete edge list of this network can be found in [10].

Second, we measured the extent to which this implicit network comply with the factors often used to classify projects. For this study we selected five factors that are often cited by popular forges (e.g., SourceForge [11]) for categorizing FLOSS projects. These factors include programming language, project size, license, project rating [4] and project domain. Project size was further categorized

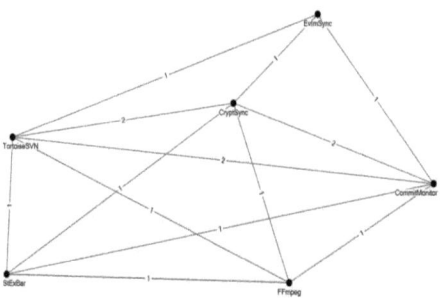

Fig. 3. An illustration of the implicit network

into very large (500K SLOC), large (50K-500K SLOC), medium (5K-50K SLOC) and small (<5K SLOC), according to current literature [19]. Similarly, project rating was classified into top (≤ 4), high (≤ 3 and <4), medium (≤ 2 and <3) and low (<2) on a scale of 5.

For each factor, we identified from the implicit network the number of edges in which both projects have the same value. Due to the large size of the implicit network, we limited this investigation to top ranked 262 edges (the edges that have weight greater than 10) and least ranked edges (random selection of 500 edges from the edges that have weight of 1).

The result of this analysis is reported in Table 2. As an illustration of this approach, consider the project factor *Language* in Table 2. Among the top 262 edges, projects in 228 edges (87.03%) use same programming language, whereas among the bottom 495 edges, projects in 233 edges (47%) have same languages.

2.5 Program Verification

Two pass evaluations were conducted to verify the correctness of the implemented programs. First, the programs were tested with limited number of data samples taken from the collected data. Notified bugs (e.g., errors in parsed data for an HTML tag) were fixed accordingly. Second, a manual checking on a random sample of the actual collected data was done. The correctness of collected data in the second pass was reported to be over 98%.

3 Result Analysis

In this section we investigate the research questions primarily based on the implicit network (described in Section 2.4) revealing projects relationships based on common contributor(s), and by evaluating its compliance with the project factors (presented in Table 2) often used to classify FLOSS projects.

1. How do FLOSS project development communities overlap?

In this study we examined the implicit relationships among 2641 FLOSS projects that are contributed to by 530 contributors. Based on common contributor(s) as a relationship criterion, the implicit network reveals 194424 edges (implicit relations) between 4261 projects. Edge weight lays between 1 and 41, which simply reflects the total number of common contributors between projects. A partial snapshot of this network is shown in Figure 4, in which, for instance, projects *Debian* and *x.Org* have a relationship edge with weight 17. This is because 17 contributors contributed to both projects. This result, all together, portrays the collaborative nature of contribution by FLOSS community members.

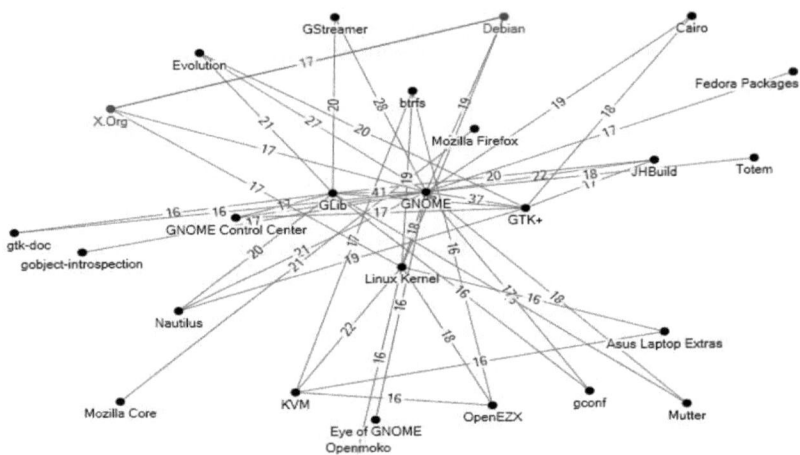

Fig. 4. Partial snapshot of the Implicit Network

To dig further and reason about such large deviation of edge weight, we counted the edges within a certain weight range, result of which is shown in Table 1. As presented in the table, the majority of the edges has low edge weight count (192559 edges out of 194424 have weight bellow 5). Investigating the cause, we observed that projects in these relationships are either medium or small size projects. Even in case where both projects have similar project sizes (3rd row of Table 2), 30% of the projects are medium or small sized. Hence, it is reasonable that communities of such projects should be small. Contrary to this, projects that are within the high edge weight count (e.g., edge weight over 10), are among the very large or large project groups, thus justifying the large overlapping community of contributors.

The above observation is analogous to the *richer gets rich* phenomenon in FLOSS projects collaboration [12], which states that communities that already had a high population would effectively attract more contributors.

Additionally, this structure of sharing contributors among multiple projects is supported by the small-world phenomenon [13]. In a small-world structure several projects are connected with each other through one or more links, e.g., common contributors. In this setup, with increasing number of common contributors, the communities of related projects (as realized by the implicit network) become strongly interconnected. We argue that this in turn may affect project success: The productivity of the contributors is boosted by providing them a dense communication channel to acquire more quantity and variety of information and knowledge resources [14].

Table 1. Edge count under edge weight category

Edge Weight Category	Edge Count
Less than 5	192559
Between 5 and 10	1603
Between 10 and 20	252
Greater than 20	10

Furthermore, contributors often participated in projects that belong to the same domain or are sub-projects to a larger one, and that utilizes same programming language(s) as development medium. High percentage of commonalities in implicit network under these two categories (as reported in row 6 and 2 of Table 2) vindicated the claim. This complies with the fact that contributors develop relationships on the common ground of interest [15].

> Based on the above observation and discussion it can be affirmed that FLOSS communities often prefer to participate in related projects with participation count varies with the size of the projects.

2. To what extent can developer sharing in FLOSS projects approximate resemblance between the projects themselves?

Within the scope of this investigation, we rationalized the projects relationship in implicit network against the actual factors that relate FLOSS projects. In doing so, we measured the extent to which the implicit network comply with the factors often used to classify projects. This approach is explained in detail in Section 2.4, and the result of which is presented in Table 2.

Among the five project factors, implicit network could effectively approximate three of them, namely, project domain, programming language and project size. Column six in Table 2 shows high percentage of compliance of the implicit relationships to these project factors.

Contributors are most often attracted towards projects that fall within the same project domain or are the sub-projects of a larger one. As can be seen in row six of Table 2, among the top listed 262 edges, 257 (98%) has conformance to same project domain. Similar observation holds (with 70% of conformance) for bottom 500 edges as well. This implies that similar project domain most effectively creates favorable ground for attracting contributors to participate in them.

Table 2. Compliance of the edges in Implicit Network to that of project factors

Project Factor	Edge Selection Category	Selected edges within the category	No of Edges in which both nodes have attribute value	Edges having same attribute value for the nodes	(%) count	Additional Info
Language	Top [Edge weight >10]	262	262	228	87.03%	
	Bottom [Edge weight = 1]	500	495	233	47%	
Project Size	Top [Edge weight >10]	262	262	168	64.13%	Very large: 139 Large: 29
	Bottom [Edge weight = 1]	500	500	150	30%	Very large: 25 Large: 70 medium: 35 Small: 20
License	Top [Edge weight >10]	262	214	84	39.26%	
	Bottom [Edge weight = 1]	500	290	96	33.1%	
Project Rating	Top [Edge weight >10]	262	182	124	68.14%	Top: 107 High: 17
	Bottom [Edge weight = 1]	500	145	130	89.66%	Top: 120, High: 10
Project Domain	Top [Edge weight >10]	262	262	257	98%	
	Bottom [Edge weight = 1]	500	500	350	70%	

Language similarity is found to be one of the major selection factors for contributors participation. According to the data in the second row of Table 2, 87.03% of the top ranked 262 edges have language similarity in contrast to only 47% similarity for the bottom 500 edges. This observation approves that language similarity among projects offers strong support to attract large number of contributors.

The factor project size also imitate analogous results to that of language factor (row three in Table 2). Projects that are very large or large in size (64.13%) are able to manage larger collaborative contributor community than medium or smaller sized projects (30%).

Additionally, results on project rating show that contributors are more attracted towards highly rated projects than low rated ones (row 5 of Table 2).

However, contributors participation to the projects is not constrained by the licenses of the respective projects. Our investigation reported that very low percentage of projects in implicit network have same license terms (only 39.26% of the top ranked edges and 33.1% of the bottom edges as presented in row 4 of Table 2).

> Projects that are linked in the implicit network with high edge weight count most likely belong to the same domain, use the same programming languages, or have similar project size.

The following aspects have been identified which could lead to threats to validity of this study.

External validity (how results can be generalized): This study includes 4261 FLOSS projects that are contributed by 530 contributors. Though, these projects cover a wide spectrum of FLOSS territory according to project size, domain, used languages and licenses, we cannot claim completeness of this justification.

Internal validity (confounding factors can influence the findings): The data used in this study is limited to the one provided by Ohloh and may raise trust concerns.

Construct validity (relationship between theory and observation): Data analysis programs written for this study produce data accuracy of over 98%, which was measured with random sample of collected data. This may affect the construct validity.

4 Conclusions

This paper studied to what extent resembling FLOSS components can be tracked based on community activities. Based on our findings, we claim that the proposed approach could approximate to a satisfactory degree resemblance between projects with respect to project domain, programming language and project size. Contrary to these factors, license terms of the projects came out as the least influential factor among all. This finding points out the fact that individual contributors may not be concerned about the licensing issues while selecting projects for contribution. These claims however, need further study. In this regard, a questionnaire to the open source community could be planned and carried out.

Acknowledgement. This work is funded by the Nokia Foundation Grant, 2013 and the TiSE Graduate school, Finland.

References

1. Crowston, K., Howison, J.: The social structure of Free and Open Source Software development. First Monday 10(2) (2005)
2. Aaltonen, T., Jokinen, J.: Influence in the Linux Kernel Community. In: Feller, J., Scacchi, B.F.W., Sillitti, A. (eds.) Open Source Development, Adoption and Innovation. IFIP, vol. 234, pp. 203–208. Springer, Boston (2007)
3. Bonaccorsi, A., Rossi, C.: Altruistic individuals, selfish firms? the structure of motivation in open source software. First Monday (1-5) (2004)
4. Ohloh, http://www.ohloh.net/ (last accessed November 2013)
5. Ohloh kudo rank, http://meta.ohloh.net/kudos/ (last accessed November 2013)
6. Deshpande, A., Riehle, D.: The Total Growth of Open Source. In: Russo, B., Damiani, E., Hissam, S., Lundell, B., Succi, G. (eds.) Open Source Development, Communities and Quality. IFIP, vol. 275, pp. 197–209. Springer, Boston (2008)

7. Rudzki, J., Kiviluoma, K., Poikonen, T., Hammouda, I.: Evaluating Quality of Open Source Components for Reuse-Intensive Commercial Solutions. In: Proceedings of EUROMICRO-SEAA 2009, pp. 11–19 (2009)
8. Apache POI-Java API for Microsoft Documents, http://poi.apache.org/ (last accessed September 2013)
9. jsoup: Java HTML Parser, http://jsoup.org/
10. Research Data, http://datasourceresearch.weebly.com/
11. Source Forge, http://sourceforge.net (last accessed November 2013)
12. Weiss, M., Moroiu, G., Zhao, P.: Evolution of Open Source Communities. In: Damiani, E., Fitzgerald, B., Scacchi, W., Scotto, M., Succi, G. (eds.) Open Source Systems. IFIP, vol. 203, pp. 21–32. Springer, Boston (2006)
13. Vir Singh, P.: The Small-World Effect: The Influence of Macro-Level Properties of Developer Collaboration Networks on Open-Source Project Success. ACM TOSEM 20(2), Article 6 (2010)
14. Watta, D.: Networks, dynamics, and the small world phenomenon. Amer. J. Sociology 105, 493–527 (1999)
15. Madey, G., Freeh, V., Tynan, R.: The open source software development phenomenon: An analysis based on social network theory. In: Americas Conf. on Information Systems, pp. 1806–1813 (2002)
16. Robles, G., González-Barahona, J.M.: A Comprehensive Study of Software Forks: Dates, Reasons and Outcomes. In: Hammouda, I., Lundell, B., Mikkonen, T., Scacchi, W. (eds.) OSS 2012. IFIP AICT, vol. 378, pp. 1–14. Springer, Heidelberg (2012)
17. Orsila, H., Geldenhuys, J., Ruokonen, A., Hammouda, I.: Update Propagation Practices in Highly Reusable Open Source Components. In: Russo, B., Damiani, E., Hissam, S., Lundell, B., Succi, G. (eds.) Open Source Systems. IFIP, vol. 275, pp. 159–170. Springer, Boston (2008)
18. Gonzalez-Barahona, J.M., Robles, G., Dueñas, S.: Collecting Data About FLOSS Development: The FLOSSMetrics Experience. In: Proceedings of FLOSS 2010, Cape Town, South Africa, pp. 29–34 (2010)
19. Scacchi, W.: Understanding Open Source Software Evolution: Applying, Breaking, and Rethinking the Laws of Software Evolution. John Wiley and Sons (2003)

How Do Social Interaction Networks Influence Peer Impressions Formation? A Case Study

Amiangshu Bosu and Jeffrey C. Carver

University of Alabama, Tuscaloosa, AL, USA
asbosu@ua.edu, carver@cs.ua.edu

Abstract. Due to their lack of physical interaction, Free and Open Source Software (FOSS) participants form impressions of their teammates largely based on sociotechnical mechanisms including: code commits, code reviews, mailing-lists, and bug comments. These mechanisms may have different effects on peer impression formation. This paper describes a social network analysis of the WikiMedia project to determine which type of interaction has the most favorable characteristics for impressions formation. The results suggest that due to lower centralization, high interactivity, and high degree of interactions between participants, the code review interactions have the most favorable characteristics to support impression formation among FOSS participants.

Keywords: Open Source, OSS, FOSS, social network analysis, collaboration.

1 Introduction

Impression Formation (i.e., obtaining an accurate perception of teammates' abilities) is difficult for Free Open Source Software (FOSS) developers because of the lack of physical interaction with their distributed, virtual teammates. Inaccurate perceptions of teammates' abilities and expertise may reduce the team's productivity because some developers improperly disregard the opinions or contributions of certain teammates. Due to the lack of physical interaction, members of distributed teams form impressions based on tasks [16]. This task-focus increases the potential effects of *socio-technical interactions* (i.e. social interactions facilitated by technological solutions) on impression formation. Common FOSS socio-technical interaction mechanisms include: project mailing-lists, bug repository, code review repository, and code repository.

Each of these socio-technical mechanisms facilitates different types of interactions among project participants. The results of our recent large-scale survey of FOSS developers suggest that developers believe code review is beneficial in the impression formation process [6]. Our hypothesis in this work is that due to the nature of interaction supported by each of these mediums, some may be more beneficial than others for impression formation. We conducted a social network analysis (SNA) of the interactions facilitated by each socio-technical mechanism to determine which of those mechanisms tended to support impression formation.

A social network is a theoretical construct that represents the relationships between individuals, organizations, or societies. These networks are typically modeled as a graph in which vertices represent individuals or organizations and edges represent a relationship between individuals or organizations. SNA focuses on understanding the patterns of interactions and the relative positions of individuals in a social settings [10]. SNA is commonly used in many domains, e.g., sociology, biology, anthropology, communication studies, information science, organizational studies, and psychology.

Here we summarize the results of previous SNA analysis on interactions facilitated by socio-technical mechanisms. *Code-commit* social networks: 1) exhibit 'small world' characteristics[1], 2) show that developers who work on the same file are more likely to interact in the mailing-list [4], and 3) identify the influence of a sponsoring company on the development process, release management, and leadership turnover [15]. *Bug-fixing* social networks show: 1) decentralized interactions in larger projects [8] and 2) a stable core-periphery structure with decreasing group centralization over time [14]. *Mailing-list* social networks: 1) exhibit 'small world' characteristics with developers having higher status [3] and 2) the core members have disproportionately large share of communication with the periphery members [17]. *Code-review* social networks show those responsible for testing and approving patches are the central actors [19].

Previous studies applying SNA to FOSS projects focused on only one or two of these social networks. However, due to the differing characteristics of each network, it is important to compare the networks to understand the various types of interactions within FOSS communities. Therefore this study has two primary goals: 1) Identify the key characteristics of each FOSS social interaction networks, and 2) Compare those networks to identify which have characteristics that best support peer impression formation.

The remainder of the paper is organized as following. Section 2 describes the research method. Section 3 defines the metrics used in this study. Section 4 discusses the study results. Finally, Section 5 concludes the paper.

2 Research Method

We performed a case study to analyze four types of social networks. This section details the methods for project selection, data collection, and data analysis.

2.1 Project Selection

To allow for the desired analysis, we needed to identify a FOSS project with a large, active developer community who performed code reviews that had publicly accessible repositories. The FOSS projects maintained by the WikiMedia foundation (www.wikimedia.org) satisfy the criteria. WikiMedia foundation is a non-profit organization that maintains many active, open-source, web-based

---

[1] The average distance between pairs of nodes in a large empirical networks are often much shorter than in random graphs of the same size [18].

projects including: MediaWiki, and Wikdata. Contributors to these projects include a good mix of volunteers and paid Wikimedia employees. This characteristic is representative of many popular FOSS projects and therefore makes these projects ideal for this research.

2.2 Data Collection

The data from WikiMedia required to conduct SNA on the four sociotechnical mechanisms is available in online repositories. The WikiMedia project development policy[2] requires all code changes to be reviewed prior to inclusion in the project. To facilitate this review process, the WikiMedia projects use Gerrit[3], an online code review tool. Because all changes were submitted to Gerrit, this repository contains the code review and the code commit information. WikiMedia's BugZilla[4] repository contains the fault data. We developed Java applications to mine these repositories. WikiMedia's mailing-list archives (i.e. wikitech-l[5]) contains the development mailing-list data. We used the Mailing List Stats tool [13] to mine the development mailing-list data. All tools populated a local MySQL database.

2.3 Data Analysis

We cleaned the data by removing 'bot' accounts from the code review and bug repositories. A manual inspection of the comments by accounts that contained 'bot' in the name (e.g. L10n-bot or jenkins-bot) revealed that the messages were automated rather than human-generated. We also merged multiple mailing-list accounts from the same developer using an existing approach [3] to resolve multiple email aliases belonging to same developer.

We then wrote scripts to calculate the number of interactions between developers captured by data in each repository. For each data type, we created social networks as undirected, weighted graphs where nodes represent developers and edge weights represent the quantity of interactions between the developers. Table 1 provides an overview of the characteristics of the four social interaction networks. We then used Gephi [2] for SNA and UCINet [5] for calculating network centralization scores.

3 Social Network Metrics

This section describes the commonly used SNA metrics employed in this study.

Average Degree: Average of the degrees of all the nodes in the network, where *degree* is the number of edges connected to a node. This metric quantifies the average number of teammates with which each person interacts.

---

[2] http://www.mediawiki.org/wiki/Developmentpolicy
[3] https://gerrit.wikimedia.org
[4] https://bugzilla.wikimedia.org/
[5] https://lists.wikimedia.org/mailman/listinfo/wikitech-l

Table 1. Overview of the four social interaction networks

Network	Data source	Edge weight between nodes
Mailing-list	29,038 emails	# of mutual email threads
Bug interaction	9,952 bugs	# of bugs where both have commented
Code review	90,186 requests	# of mutual code review requests
Code commit	90,186 commits	# of files where both have made at least one change

Average Weighted Degree: Average of the weighted degrees of all the nodes in the network, where *weighted degree* is the sum of the edge weights of all edges connected to a node. This metric indicates the average volume of interaction between any two teammates.

Network Density: The ratio of the number of edges that exist to the maximum number of edges possible. This metric indicates how many pairs of teammates are interacting with each other.

Because many real-world networks exhibit small-world *properties, the next three metrics quantify the small-world characteristics of the networks.*

Average Path Length: The average length of the shortest paths (i.e. number of steps) between every pair of nodes. This metric indicates the average distance in the interaction network between any two teammates.

Average Clustering Coefficient: The *clustering coefficient* of a node is a measure of the connectedness of its neighbors. That is, it is the percentage of all possible edges between a node's neighbors that are actually present or the probability that any two neighbors of the node are connected. The *Average Clustering Coefficient* of a network is the average of the clustering coefficients of the nodes in the graph. Small-world networks have a higher average clustering coefficient relative to other networks of same size [18].

Network Centralization: Network centralization is a measure of the relative centrality of the most central nodes in a network [9]. A higher centralization means that there are a relatively small number of central nodes. According to Freeman [9], network centralization can be calculated as

$$C_X = \frac{\sum_{i=1}^{n}[C_X(p*)-C_X(p_i)]}{max \sum_{i=1}^{n}[C_X(p*)-C_X(p_i)]}$$

where n= number of nodes in the network, $C_X(p_i)$ = any centrality measure of point i, $C_X(p*)$ = largest value of $C_X(p_i)$ for any node in the network, and $max \sum_{i=1}^{n}[C_X(p*)-C_X(p_i)]$= the maximum possible sum of differences in point centrality for a network on n nodes. Among the different methods for calculating network centrality, Freeman recommends using *Betweenness Centrality* (i.e. the number of times a node acts as a bridge along the shortest path between two other nodes [9]), to identify the central nodes in a network [11]. Therefore, we used *Betweenness Centrality* as an indicator of the extent to which a small number of central nodes dominate the interactions.

4 Results

Figure 1 shows the social networks built from the four data sources using the Fruchterman-Reingold layout algorithm [12]. To make the diagrams more comprehensible, we first filtered out the low-weight edges and then excluded the isolated nodes. Node size is based on weighted degrees. Edge widths are based on edge weights. Intensity of the node color is based on betweenness centrality.

A visual inspection of the diagrams suggests that the *bug interaction* and *mailing-list* networks are similar. They each contain only a few nodes with high weighted degree (i.e. node size) and high betweenness centrality (i.e. color intensity). There are also a small number of edges between non-central nodes, which indicates little interaction among them. Conversely, the *code review* and *code commit networks* have more nodes with high weighted degree and high betweenness centrality. Furthermore, in the *code commit network*, the distribution of nodes based on weighted degree and betweenness centrality is more uniform and contains a lot of edges between the non-central nodes. These graphs suggest that the *mailing-list* and *bug interactions* networks are the most centralized, followed by the *code review* network and finally the *code commit* network, which is the most decentralized.

The following subsections provide a more detailed comparison of the networks using the metrics defined in Table 2.

Table 2. Metrics describing each social interaction network

	Mailing-list	Bug interaction	Code review	Code commit
# of nodes	2,518	3,416	522	448
# of edges	28,208	34,200	5,052	21,353
Avg. degree	22.33	20.02	19.36	95.33
Avg. weighted degree	28.35	39.12	196.93	688.55
Avg. path length	2.67	2.6	2.53	1.88
Network density	0.009	0.006	0.037	0.213
Avg. clustering coefficient	0.762	0.853	0.693	0.839
Network centralization	28.49%	32.18%	10.94%	5.94%

4.1 Network Size

The *bug interaction* network has the most nodes, followed by the *mailing-list* network, the *code review* network and finally the *code commit* network. The *bug interaction network* is the largest because any user can create an account and post bugs in the BugZilla repository, while the development *mailing-list* is only for development discussion and posting requires approval from the list administrator. The *code review* network is slightly larger than the *code commit* network because a contributor without commit access can submit a patch for code review. If that patch passes the code review, a core developer commits it to the main branch. Finally, because only long time contributors earn commit access, the *code commit* network is the smallest.

Participants have a similar number of peers in all networks except for the *code commit* network, where they have more. This larger number of peers may indicate that developers work on many modules. Although participants in the *mailing-list*, *bug*, and *code review* networks have a similar number of peers, participants

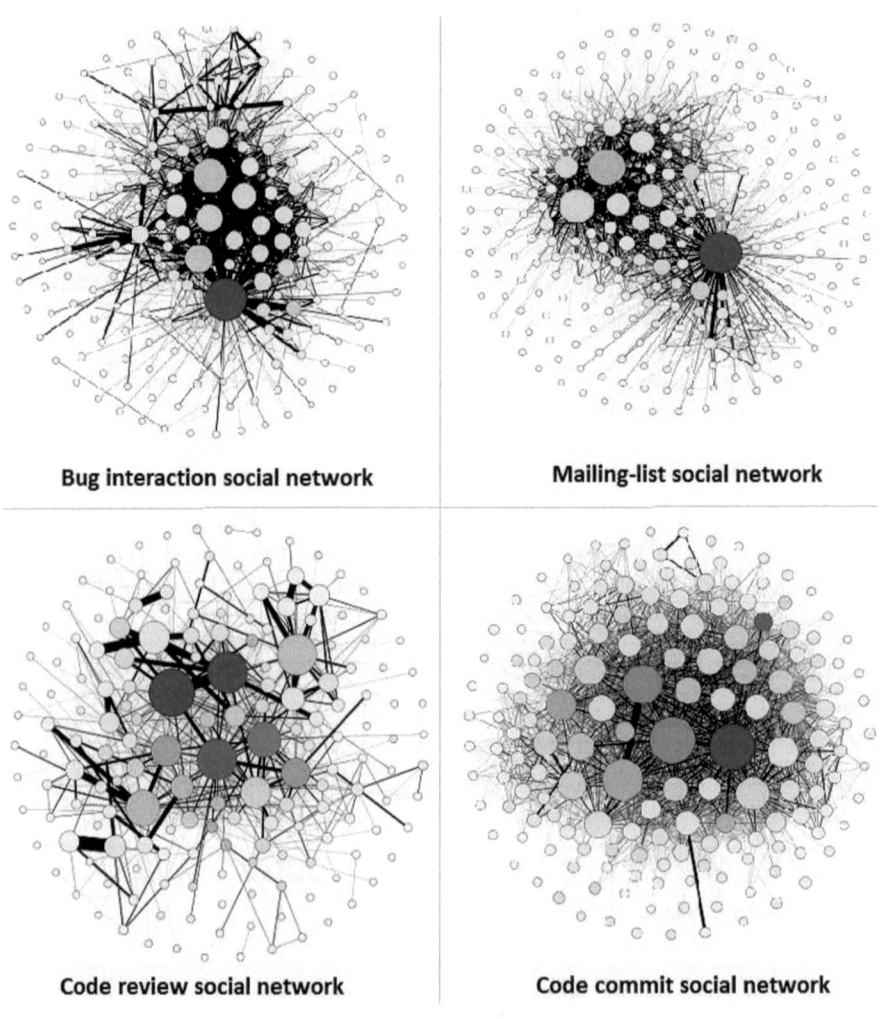

Fig. 1. Social network diagrams

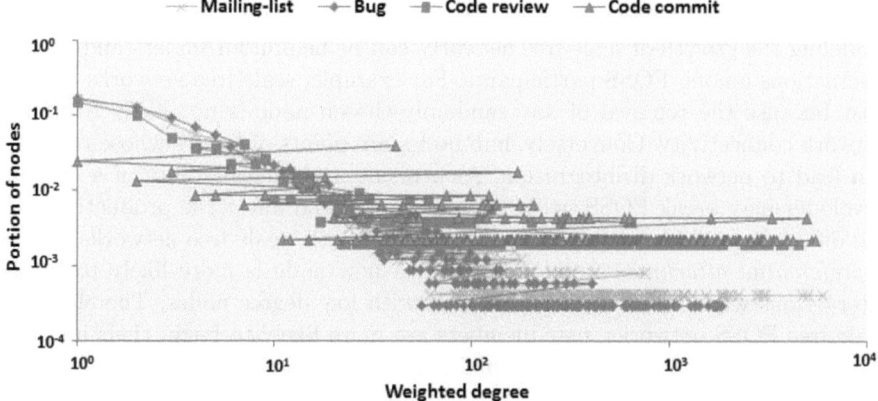

Fig. 2. Weighted degree distributions (log-log)

in the *code review* network interact more frequently with those peers (i.e. higher average weighted degree). Because of the very high average degree, the *code commit* network has the lowest average path length, and highest density. The *mailing-list* and *bug interaction* networks have similar densities and average path lengths. However, the *code review* network has a lower average path length, and higher density, indicating that two contributors are more likely to interact through code review than through a mailing-list or a bug. Considering the higher probability (i.e. density) and higher volume (i.e weighted degree) of *code commit* and *code review* interactions, developers are more likely to form impressions with their teammates via those interactions.

4.2 Network Model

Figure 2 shows that for three of the four networks (excluding the code commit network), the weighted degree distribution may follow the power law distribution $(p[X = x] \propto (x)^{-\alpha})$[6]. We used the R - poweRlaw package [7] to estimate the α values (i.e. power law exponent) and the goodness-of-fit[7] (p) for the power law distributions. The α values for the *mailing-list*, *bug interaction*, and *code review* networks are estimated as 1.73 (p=.976), 1.7 (p=.994), and 1.9 (p=.851), respectively. This result suggests that like many other real world social networks, the *mailing-list*, *bug interaction*, and *code review* networks follow the power law distribution. The low value of $\alpha (< 3)$, short average path length (Table 2), 'small world' property (i.e. high clustering coefficients), and presence of hubs that indicate those three graphs follow the *scale-free network* model [1].

---

[6] In a power law distribution, small occurrences are extremely common and large instances are extremely rare. Because the probability of a large value decreases exponentially, this disribution is referred to as the 'power law'.

[7] Kolmogorov-Smirnov goodness of fit test measures if the dataset comes from a specific distribution. p $\approx$ 0 indicates the model does not provide a plausible fit.

The results of previous research [1] into understanding the properties and modeling the growth of scale-free networks can be helpful for understanding the interactions among FOSS participants. For example, scale-free networks are robust because the removal of any randomly-chosen node is not likely to break network connectivity. Conversely, hub nodes are points of failure whose removal can lead to network disintegration. As a result, the unavailability of a central developer may break FOSS network connectivity and affect the productivity of the other dependent developers. In terms of growth, scale-free networks follow a *preferential attachment* model in which a new node is more likely to begin interactions with high degree nodes than with low degree nodes. Therefore, in scale-free FOSS networks, new members are more likely to begin their interactions with contributors who are popular (i.e. the higher degree nodes).

4.3 Network Centralization

The betweenness centralization scores indicate that the *bug interaction* and the *mailing-list* networks are the most centralized. They depend more upon the central nodes than do the *code commit* and *code review* networks.

High centralization might be beneficial in some cases (i.e. making quick decisions). But for a large FOSS network, high centralization has many drawbacks. First, central nodes may become bottlenecks due to overload. Second, because most interaction goes through the central hubs, there is less interaction among the non-central nodes. In this situation, the central nodes become familiar with most of the network while the non-central nodes become familiar only with the hubs. Therefore, high centralization is not helpful for impression formation in a FOSS community because most nodes are non-central nodes. As a result, lower centralization may be better for impressions formation.

Because it has the lowest centralization score, the *code commit* network has the most favorable characteristic for impression formation. While working on code written by a teammate could lead to impression formation, the lack of the type of direct interaction that occurs during code review suggests that interactions around code commits may not be the best for impression formation. The *code review* network has the second lowest centralization score. As opposed to the code commit interactions, code review interactions are highly interactive between the author and reviewer. Evaluating the quality of code written by a peer and discussing it's design helps to build mutual impressions between the author and reviewers. Although mailing-lists and bug interaction are also interactive, those interactions are highly centralized.

Therefore, considering both interactivity and centralization, we hypothesize that code review interactions should provide the best support for impression formation between FOSS developers. The results of our prior study, which indicate a high level of trust, reliability, perception of expertise, and friendship between FOSS peers who have participated in code review for a period of time, also provide evidence for this hypothesis [6].

5 Conclusion

Considering network structure, centralization, and number of interactions, interactions related to code commits have the most favorable characteristics for impression formation, followed by interactions related to code review. However, because the interactions related to code commits lack the type of direct interactivity present in those related to code review, we hypothesize that the *code review activity may be best suited to support impression formation*. Despite the differences in SNA metric values, we hypothesize that the code review and code commit interactions are complementary for impressions formation, because developers who work on the same module are more likely to review each others' code. While in this paper we did not study this hypothesis, we consider testing this hypothesis as a future goal.

Additionally, the characteristics of the four WikiMedia social networks are quite different, except for some similarities in the *mailing-list* and *bug interaction* networks. These different characteristics suggest that each social network provides a different perspective on interactions within an FOSS community. To fully understand the community interactions and the peer impression formation process, it is likely necessary to examine multiple perspectives rather than drawing a conclusion based only on one type of interaction.

The primary threats to validity for this are related to external validity. Due to the vast differences among FOSS projects, the results of this one case study are not generalizable to the entire FOSS domain. To build reliable empirical evidence, we need to study more FOSS projects. We carefully selected WikiMedia projects for this case study as those might be good representatives of community-driven FOSS projects. The results of this study can help to build research questions for future FOSS studies.

The contributions of this paper include:

- Identified the need for studying different interaction networks to understand FOSS community structure;
- Identified the scale-free network model to describe FOSS participants' interaction networks; and
- Identified the code review interaction as a likely method for impression formation among FOSS participants.

The future work for this research includes replicating this case study using more FOSS projects to verify the generalizability of this result.

Acknowledgment. This research is supported by US National Science Foundation grant 1322276.

References

1. Barabási, A.L., Albert, R.: Emergence of scaling in random networks. Science 286(5439), 509–512 (1999)
2. Bastian, M., Heymann, S., Jacomy, M.: Gephi: An open source software for exploring and manipulating networks. In: Proc. 3rd Int'l. AAAI Conf. on Weblogs and Social Media, San Jose, California (2009)
3. Bird, C., Gourley, A., Devanbu, P., Gertz, M., Swaminathan, A.: Mining email social networks. In: Proc. 3rd Int'l Wksp. on Mining Soft. Repositories, Shanghai, China, pp. 137–143 (2006)
4. Bird, C., Pattison, D., D'Souza, R., Filkov, V., Devanbu, P.: Latent social structure in open source projects. In: Proc. 16th ACM SIGSOFT Int'l Symp. on Foundations of Soft. Eng., Atlanta, Georgia, pp. 24–35 (2008)
5. Borgatti, S.P., Everett, M.G., Freeman, L.C.: Ucinet for windows: Software for social network analysis (2002)
6. Bosu, A., Carver, J.C.: Impact of peer code review on peer impression formation: A survey. In: Proc. 7th ACM/IEEE Int'l. Symp. on Empirical Soft. Eng. and Measurement, Baltimore, MD, USA, pp. 133–142 (2013)
7. Clauset, A., Shalizi, C., Newman, M.: Power-law distributions in empirical data. SIAM Review 51(4), 661–703 (2009)
8. Crowston, K., Howison, J.: The social structure of free and open source software development. First Monday 10(2-7) (2005)
9. Freeman, L.C.: Centrality in social networks conceptual clarification. Social Networks 1(3), 215–239 (1979)
10. Freeman, L.C.: The development of social network analysis: A study in the sociology of science, vol. 1. Empirical Press, Vancouver (2004)
11. Freeman, L.C., Roeder, D., Mulholland, R.R.: Centrality in social networks: II. experimental results. Social Networks 2(2), 119–141 (1980)
12. Fruchterman, T.M., Reingold, E.M.: Graph drawing by force-directed placement. Software: Practice and Experience 21(11), 1129–1164 (1991)
13. Herraiz, I., Perez, J.G.: Mailing list stats
14. Long, Y., Siau, K.: Social network structures in open source software development teams. Journal of Database Mgmt. 18(2), 25–40 (2007)
15. Martinez-Romo, J., Robles, G., Gonzalez-Barahona, J.M., Ortuño-Perez, M.: Using social network analysis techniques to study collaboration between a FLOSS community and a company. In: Russo, B., Damiani, E., Hissam, S., Lundell, B., Succi, G. (eds.) Open Source Development, Communities and Quality. IFIP, vol. 275, pp. 171–186. Springer, Boston (2008)
16. McKenna, K.Y., Bargh, J.A.: Plan 9 from cyberspace: The implications of the internet for personality and social psychology. Personality and Social Psychology Review 4(1), 57–75 (2000)
17. Oezbek, C., Prechelt, L., Thiel, F.: The onion has cancer: Some social network analysis visualizations of open source project communication. In: Proc. 3rd Int'l. Wksp. on Emerging Trends in Free/Libre/Open Source Soft. Research and Development, Cape Town, South Africa, pp. 5–10 (2010)
18. Watts, D.J., Strogatz, S.H.: Collective dynamics of 'small-world' networks. Nature 393(6684), 440–442 (1998)
19. Yang, X., Kula, R.G., Erika, C.C.A., Yoshida, N., Hamasaki, K., Fujiwara, K., Iida, H.: Understanding oss peer review roles in peer review social network (PeRSoN). In: Proc. 19th Asia-Pacific Soft. Eng. Conf., Hong Kong, pp. 709–712 (2012)

Drawing the Big Picture: Temporal Visualization of Dynamic Collaboration Graphs of OSS Software Forks

Amir Azarbakht and Carlos Jensen

Oregon State University, School of Electrical Engineering & Computer Science,
1148 Kelley Engineering Center, Corvallis OR 97331, USA
{azarbaam,cjensen}@eecs.oregonstate.edu

Abstract. How can we understand FOSS collaboration better? Can social issues that emerge be identified and addressed as they happen? Can the community heal itself, become more transparent and inclusive, and promote diversity? We propose a technique to address these issues by quantitative analysis and temporal visualization of social dynamics in FOSS communities. We used social network analysis metrics to identify growth patterns and unhealthy dynamics; This gives the community a heads-up when they can still take action to ensure the sustainability of the project.

1 Introduction

Social networks are a ubiquitous part of our social lives, and the creation of online social communities has been a natural extension of this phenomena. Free/Open Source Software (FOSS) development efforts are prime examples of how community can be leveraged in software development, groups are formed around communities of interest, and depend on continued interest and involvement in order to stay alive [17].

Though the bulk of collaboration and communication in FOSS communities occurs online and is publicly accessible, there are many open questions about the social dynamics in FOSS communities. Projects might go through a metamorphosis when faced with an influx of new developers or the involvement of an outside organization. Conflicts between developers' divergent visions about the future of the project might lead to forking of the project and dilution of the community. Forking, either as a violent split when there is a conflict or as a friendly divide when new features are experimentally added both affect the community [3].

Most recent studies of FOSS communities have tended to suffer from an important limitation. They treat community as a static structure rather than a dynamic process. In this paper, we propose to use temporal social network analysis to study the evolution and social dynamics of FOSS communities. With these techniques we aim to identify measures associated with unhealthy group dynamics, e.g. a simmering conflict, as well as early indicators of major events in the lifespan of a community. One set of dynamics we are especially interested in, are those that lead FOSS projects to fork. We used the results of a study of forked FOSS projects by Robles and Gonzalez-Barahona [19] as the starting point for out study, and tried to gain a better understanding of the evolution of these communities.

This paper is organized as follows: We present related literature on online social communities. We then present the gap in the literature, and discuss why the issue needs to be addressed. After that, in methodology, we describe how gathering data, doing the analysis, and the visualization of the findings was carried out. At the end, we present results, discussion and threats to validity.

2 Related Work

The social structures of FOSS communities have been studied extensively. Researchers have studied the social structure and dynamics of team communications [4][10][11], identifying knowledge brokers and associated activities [20], project sustainability [10], forking [18] [19], their topology [4], their demographic diversity [13], gender differences in the process of joining them [12] and the role of the core team in their communities [21], etc. All of these studies have tended to look at community as a static structure rather than a dynamic process. This makes it hard to determine cause and effect, or the exact impact of social changes.

The study of communities has grown in popularity in part thanks to advances in social network analysis. From the earliest works by Zachary [22] to the more recent works of Leskovec et al. [14][15], there is a growing body of quantitative research on online communities. The earliest works on communities was done with a focus on information diffusion in a community [22]. Zachary investigated the fission of a community, the process of communities splitting into two or more parts. He found that fission could be predicted by applying the Ford-Fulkerson min-cut algorithm [6] on the group's communication graph; "the unequal flow of sentiments across the ties" and discriminatory sharing of information lead to "subcommunities with more internal stability than the community as a whole."

Community splits in FOSS are referred to as forks, and are relatively common. Forking is defined as "when a part of a development community (or a third party not related to the project) starts a completely independent line of development based on the source code basis of the project." Robles and Gonzalez-Barahona [19] identified 220 significant FOSS projects that have forked over the past 30 years, and compiled a comprehensive list of the dates and reasons for forking. They classified these into six main categories. (Table 3.) which we build on extensively. They identified a gap in the literature in case of "how the community moves when a fork occurs".

The dynamic behavior of a network and identifying key events was the aim of a study by Asur et al [1]. They studied three DBLP co-authorship networks and defined the evolution of these networks as following one of these paths: a) Continue, b) k-Merge, c) k-Split, d) Form, or e) Dissolve. They also defined four possible transformation events for individual members: 1) Appear, 2) Disappear,

3) Join, and 4) Leave. They compared groups extracted from consecutive snapshots, based on the size and overlap of every pair of groups. Then, they labeled groups with events, and used these identified events. The communication patterns of FOSS developers in a bug repository were examined by Howison et al. [10]. They calculated out-degree centrality as their metric. Out-degree centrality measures the proportion of the number of times a node contacted other nodes (outgoing) over how many times it was contacted by other nodes (incoming). They calculated this centrality over time "in 90-day windows, moving the window forward 30 days at a time." They found that "while change at the center of FOSS projects is relatively uncommon," participation across the community is highly skewed, following a power-law distribution, where many participants appear for a short period of time, and a very small number of participants are at the center for long periods. Our approach is similar to theirs in how we form collaboration graphs and perform our temporal analysis. Our approach is different in terms of our project selection criteria, the metrics we examine, and our research questions.

The tension between diversity and homogeneity in a community was studied by Kunegis et al. [13]. They defined five network statistics used to examine the evolution of large-scale networks over time. They found that except for the diameter, all other measures of diversity shrunk as the networks matured over their lifespan. Kunegis et al. [13] argued that one possible reason could be that the community structure consolidates as projects mature.

Community dynamics was the focus of a recent study by Hannemann and Klamma [8] on three open source bioinformatics communities. They measured "age" of users, as starting from their first activity and found survival rates and two indicators for significant changes in the core of the community. They identified a survival rate pattern of 20-40-90%, meaning that only 20% of the newcomers survived after their first year, 40% of the survivors survived through the second year, and 90% of the remaining ones, survived over the next years. As for the change in the core, they suggested that a falling maximal betweenness in combination with an increasing network diameter as an indicator for a significant change in the core, e.g. retirement of a central person in the community. Our approach builds on top of their findings, and the evolution of betweenness centralities and network diameters for the projects in our study are depicted in the following sections.

To date, most studies on FOSS have only been carried out on a small number of projects, and using snapshots in time. To our knowledge, no study has been done of project forking that has taken into account the temporal dimension.

3 Methodology

We argue that the social interactions data reflects the changes the community goes through, and will be able to describe the context surrounding a forking event. Robles and Gonzalez-Barahona [19] classify forking into six classes, listed in Table 1, based on the motives for forking.

Table 1. The main reasons for forking as classified by Robles and Gonzalez-Barahona [19]

Reason for forking	Example forks
Technical (Addition of functionality)	Amarok & Clementine Player
More community-driven development	Asterisk & Callweaver
Differences among developer team	Kamailio & OpenSIPS
Discontinuation of the original project	Apache web server
Commercial strategy forks	LibreOffice & OpenOffice.org
Legal issues	X.Org & XFree

The first three of the six motives listed are social, and so should arguably be reflected in the social interaction data. For example, if a fork occurs because of a desire for "more community-driven development", we would perhaps expect to see patterns in the data showing a strongly-connected core that is hard to penetrate for the rest of the community prior to the fork. In other words, the power stays in the hands of the same people over a long period of time while new people come and go. Our goal was to visualize and quantify how the community is structured, how it evolves, and the degree to which involvement changes over time. To this end, we picked projects from the aforementioned three categories of forked projects. This involved obtaining communication archives, creating the collaboration graphs, applying social network analysis (SNA) techniques to measure key metrics, and visualizing the evolving graphs. We did this in four phases as described in the following:

3.1 Phase 1: Data Collection

The study of forks by Robles and Gonzalez-Barahona [19] included information on 220 forks and their reasons. We applied three selection criteria to those projects. A project was short-listed if it was recent, i.e. the fork had happened after the year 2000; data was available; and their communities were of approximately the same size. This three stage filtering process resulted in the projects listed in Table 2.

Data collection involved analyzing mailing list archives. We collected data for the year in which the fork happened, as well as for three month before and three

Table 2. Forked projects for which collaboration data was collected

Projects	Reason for forking	Year
Amarok & Clementine Player	Technical (Addition of functionality)	2010
Asterisk & Callweaver	More community-driven development	2007
Kamailio & OpenSIPS	Differences among developer team	2008

months after that year in order to capture the social context context at the time of the fork.

3.2 Phase 2: Creating Communication Graphs

Many social structures can be represented as graphs. The nodes represent actors/players and the edges represent the interaction between them. Such graphs can be a snapshot of a network – a static graph – or a changing network, also called a dynamic graph. In this phase, we processed the data to form a communication graph of the community. We were looking for how people interacted with each other. We decided to treat the general mailing list as a person, because the bulk of the communication was targeted at it, and most newcomers start by sending their questions to the general mailing list. Each communication effort was captured with a time-stamp. This allowed us to form a dynamic graph, in which the nodes would exist if and only if they had an interaction with another node during the period we were interested in.

3.3 Phase 3: Temporal Visualization and Temporal Evolution Analysis

In this phase, we wanted to analyze the changes that happen to the community over a given period of time, i.e. three months before and three months after the year in which the forking event happened. We measured betweenness centrality [5] of the most significant nodes in the graph, and the graph diameter over time. Figures 2-4 show the betweenness centralities over the 1.5 year period for the Camailio, Amarok and Asterisk projects respectively. To do temporal analysis, we had two options; 1) look at snapshots of the network state over time, (e.g. to look at the network snapshots in every week, the same way that a video is composed of many consecutive frames), and 2) look at a period through a time window. We preferred the second approach, and looked through a time window of three months wide with 1.5 month overlaps. To create the visualizations, we used a 3 months time frame that progressed six days a frame. In this way, we had a relatively smooth transition.

We visualized the dynamic network changes using Gephi [2]. The videos show how the community graph is structured, using a continuous force-directed linear-linear model, in which the nodes are positioned near or far from each other proportional to the graph distance between them. This results in a graph shape between between Früchterman & Rheingold's [7] layout and Noack's LinLog [16].

4 Results and Discussion

4.1 Kamailio Project

Figure 1 shows four key frames from the Kamailio project's social graph around the time of their fork (the events described here are easier to fully grasp by watching the video. A node's size in a proportional to the number of interactions the node (contributor) has had within the study period and the position and edges of the nodes change if they had interactions within the time window shown, with six day steps per frame. The 1 minute and 37 seconds video shows the life of the Kamailio project between October 2007, and March 2009. Nodes are colored based on the modularity of the network.

The community starts with the GeneralList as the the biggest node, and four larger core contributors and three lesser size core contributors. The big red-colored node's transitions are hard to miss, as this major contributor departs from the core to the periphery of the network (Video minute 1:02) and then leaves the community (Video minute 1:24) capturing either a conflict or retirement. This corresponds to the personal difference category of forking reasons.

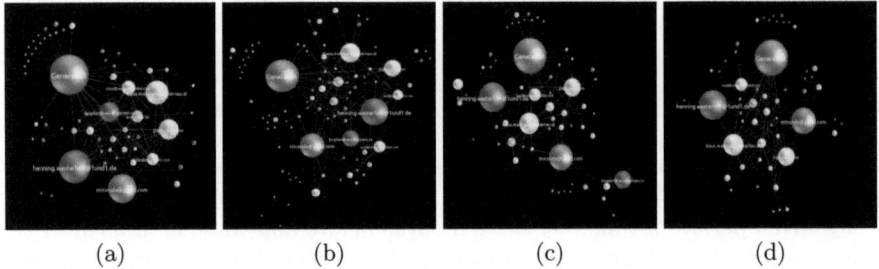

Fig. 1. Snapshots from video visualization of Kamailio's graph (Oct. 2007 - Mar. 2009) in which a core contributor (colored red) moves to the periphery and eventually departs the community

Figure 2 shows the betweenness centrality of the major contributors of Kamailio project over the same time period. The horizontal axis marks the dates, (each mark represents a 3-month time window with 1.5 months overlap). The vertical axis shows the percentage of the top betweenness centralities for each node. The saliency of the GeneralList – colored as light blue – is apparent due to to its continuous and dominant presence in the stacked area chart. The chart legend lists the contributors based on the color and in the same order of appearance on the chart starting from the bottom. One can easily see that around the "Aug. 15, 2008 - Nov. 15, 2008" tick mark on the horizontal axis, several contributors' betweenness centralities shrink to almost zero and disappear. This helps identify the date of fork with a month accuracy. The network diameter of the Kamailio project over the same time period is also shown in Figure 3. The increase in the network diameter during this period confirms the findings of Hannemann and Klamma [8].

This technique can be used to identify the people involved in conflict and the date the fork happened with a months accuracy, even if the rival project does not emerge immediately.

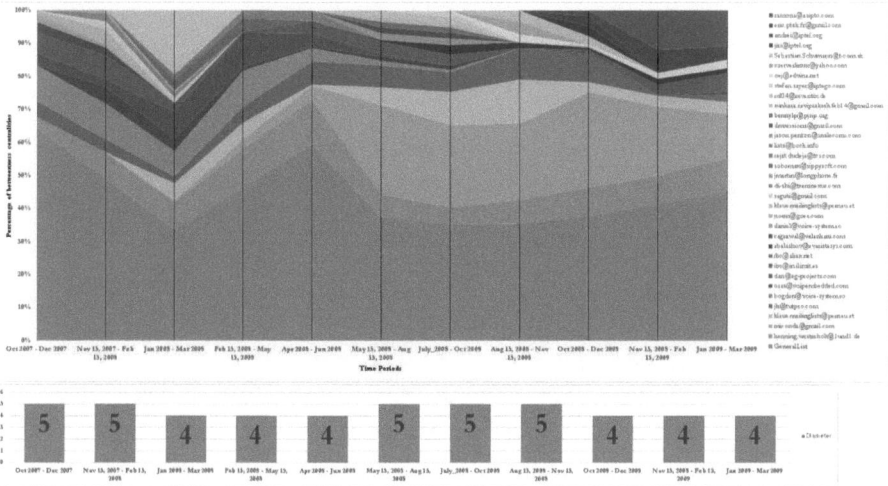

Fig. 2. Kamailio top contributors' betweenness centralities and network diameter over time (Oct. 2007 to Mar. 2009) in 3-month time windows with 1.5-month overlaps

4.2 Amarok Project

The video for the Amarok project fork is available online[1], and the results from our quantitative analysis of the betweenness centralities and the network diameters are shown in Figure 3. The results show that the network diameter has not increased over the period of the fork, which shows a resilient network. The video shows the dynamic changes in the network structure, again typical of a healthy network, rather than of simmering conflict. These indicators show that Amarok fork in 2010 arguably belongs to the "addition of technical functionality" rationale for forking, as there are no visible social conflict.

4.3 Asterisk Project

The video for the Asterisk project is also available online, and the results from our quantitative analysis of the betweenness centralities and the network diameters are shown in Figure 4. The results show that the network diameter remained steady at 6 throughout the period. The Asterisk community was by far the most crowded project, with 932 nodes and 4282 edges. The stacked area chart shows the distribution of centralities, where we see an 80%-20% distribution (, i.e. 80% or more of the activity is attributed to six major players, with the rest of the community accounting for only 20%). This is evident in the video representation as well, as the top-level structure of the network holds throughout the time period. The results from the visual and quantitative analysis links the Asterisk fork to the more community-driven category of forking reasons.

---

[1] Video visualizations available at
http://eecs.oregonstate.edu/~azarbaam/OSS2014/

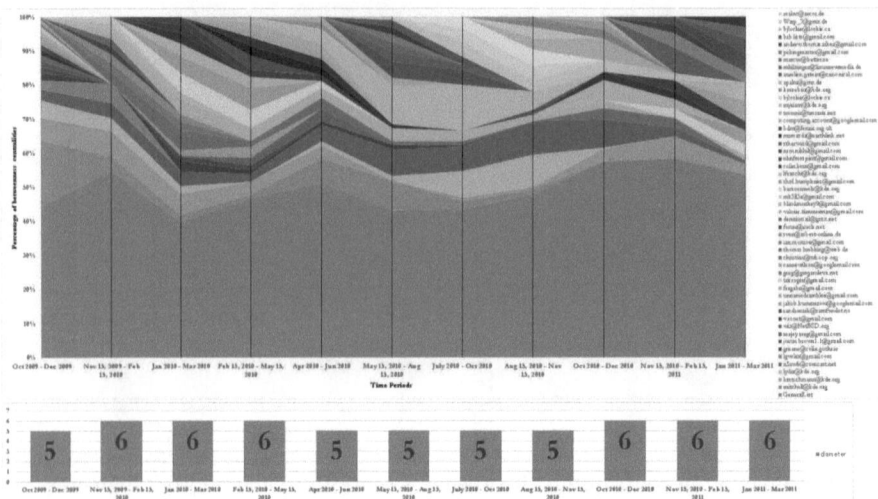

Fig. 3. Amarok project's top contributors' betweenness centralities and network diameter over time between Oct. 2009 to Mar. 2011 in 3-months time windows with 1.5 months overlaps

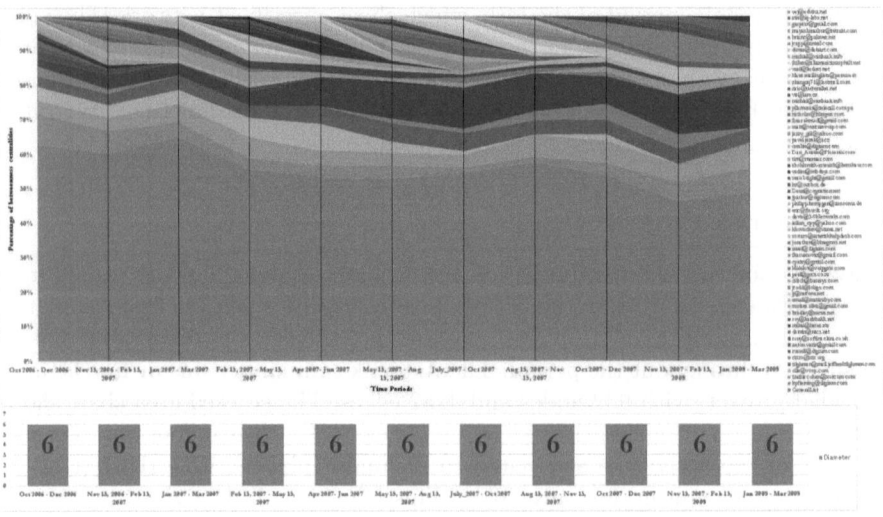

Fig. 4. Asterisk project's top contributors' betweenness centralities and network diameter over time between Oct. 2009 to Mar. 2011 in 3-months time windows with 1.5 months overlaps

5 Conclusion

We studied the collaboration networks of three FOSS projects using a combination of temporal visualization and quantitative analysis. We based our study on two papers by Robles and Gonzalez-Barahona [19] and Hannemann and Klamma [8], and identified three projects that had forked in the recent past. We mined the collaboration data, formed dynamic collaboration graphs, and measured social network analysis metrics over an 18-month period time window.

We also visualized the dynamic graph (available online) and as stacked area charts over time. The visualizations and the quantitative results showed the differences among the projects in the three forking reasons of personal differences among the developer teams, technical differences (addition of new functionality) and more community-driven development. The personal differences representative project was identifiable, and so was the date it forked, with a month accuracy. The novelty of the approach was in applying the temporal analysis rather than static analysis, and in the temporal visualization of community structure. We showed that this approach shed light on the structure of these projects and reveal information that cannot be seen otherwise.

References

1. Asur, S., Parthasarathy, S., Ucar, D.: An event-based framework for characterizing the evolutionary behavior of interaction graphs. ACM Trans. Knowledge Discovery Data 3(4), Article 16, 36 p. (2009)
2. Bastian, M., Heymann, S., Jacomy, M.: Gephi: an open source software for exploring and manipulating networks. Presented at the Int. AAAI Conf. on Weblogs and Social Media (2009)
3. Bezrukova, K, Spell, C.S., Perry, J.L.: Violent Splits Or Healthy Divides? Coping With Injustice Through Faultlines. Personnel Psychology 63(3) (2010)
4. Bird, C., Pattison, D., D'Souza, R., Filkov, V., Devanbu, P.: Latent social structure in open source projects. In: Proc. of the 16th ACM SIGSOFT Int. Symposium on Foundations of Software Engineering, pp. 24–35. ACM, New York (2008)
5. Brandes, U.: A Faster Algorithm for Betweenness Centrality. Journal of Mathematical Sociology 25(2), 163–177 (2001)
6. Ford, L.R., Folkerson, D.R.: A simple algorithm for finding maximal network flows and an application to the Hitchcock problem. Canadian Journal of Mathematics 9, 210–218 (1957)
7. Fruchterman, T.M.J., Reingold, E.M.: Graph drawing by force-directed placement. Softw: Pract. Exper. 21(11), 1129–1164 (1991)
8. Hannemann, A., Klamma, R.: Community Dynamics in Open Source Software Projects: Aging and Social Reshaping. In: Petrinja, E., Succi, G., El Ioini, N., Sillitti, A. (eds.) OSS 2013. IFIP AICT, vol. 404, pp. 80–96. Springer, Heidelberg (2013)
9. Howison, J., Crowston, K.: The perils and pitfalls of mining SourceForge. In: Proceedings of the Int. Workshop on Mining Software Repositories (MSR 2004), pp. 7–11 (2004)
10. Howison, J., Inoue, K., Crowston, K.: Social dynamics of free and open source team communications. In: Damiani, E., Fitzgerald, B., Scacchi, W., Scotto, M., Succi, G. (eds.) Open Source Systems. IFIP, vol. 203, pp. 319–330. Springer, Boston (2006)

11. Howison, J., Conklin, M., Crowston, K.: FLOSSmole: A collaborative repository for FLOSS research data and analyses. Int. Journal of Information Technology and Web Engineering 1(3), 17–26 (2006)
12. Kuechler, V., Gilbertson, C., Jensen, C.: Gender Differences in Early Free and Open Source Software Joining Process. In: Hammouda, I., Lundell, B., Mikkonen, T., Scacchi, W. (eds.) OSS 2012. IFIP AICT, vol. 378, pp. 78–93. Springer, Heidelberg (2012)
13. Kunegis, J., Sizov, S., Schwagereit, F., Fay, D.: Diversity dynamics in online networks. In: Proc. of the 23rd ACM Conf. on Hypertext and Social Media, USA (2012)
14. Leskovec, J., Kleinberg, J., Faloutsos, C.: Graphs over time: densification laws, shrinking diameters and possible explanations. In: Proc. of the SIGKDD Int. Conf. on Knowledge Discovery and Data Mining (2005)
15. Leskovec, J., Lang, K.J., Dasgupta, A., Mahoney, M.W.: Statistical properties of community structure in large social and information networks. In: Proc. of the 17th Int. Conf. on World Wide Web (WWW 2008). ACM (2008)
16. Noack, A.: Energy models for graph clustering. J. Graph Algorithms Appl. 11(2), 453–480 (2007)
17. Nyman, L.: Understanding code forking in open source software. In: Proc. of the 7th Int. Conf. on Open Source Systems Doctoral Consortium, Salvador, Brazil (2011)
18. Nyman, L., Mikkonen, T., Lindman, J., Fougère, M.: Forking: the invisible hand of sustainability in open source software. In: Proc. of SOS 2011: Towards Sustainable Open Source (2011)
19. Robles, G., González-Barahona, J.M.: A comprehensive study of software forks: Dates, reasons and outcomes. In: Hammouda, I., Lundell, B., Mikkonen, T., Scacchi, W. (eds.) OSS 2012. IFIP AICT, vol. 378, pp. 1–14. Springer, Heidelberg (2012)
20. Sowe, S., Stamelos, L., Angelis, L.: Identifying knowledge brokers that yield software engineering knowledge in OSS projects. Information and Software Technology 48, 1025–1033 (2006)
21. Torres, M.R.M., Toral, S.L., Perales, M., Barrero, F.: Analysis of the Core Team Role in Open Source Communities. In: 2011 Int. Conf. on Complex, Intelligent and Software Intensive Systems (CISIS), pp. 109–114. IEEE (2011)
22. Zachary, W.: An information flow model for conflict and fission in small groups. Journal of Anthropological Research 33(4), 452–473 (1977)

Analyzing the Relationship between the License of Packages and Their Files in Free and Open Source Software

Yuki Manabe[1], Daniel M. German[2,3], and Katsuro Inoue[3]

[1] Kumamoto University, Japan
[2] University of Victoria, Canada
[3] Osaka University, Japan

Abstract. Free and Open Source Software (FOSS) is widely reused today. To reuse FOSS one must accept the conditions imposed by the software license under which the component is made available. This is complicated by the fact that often FOSS packages contain files from many licenses. In this paper we analyze the source code of packages in the Fedora Core Linux distribution with the goal of discovering the relationship between the license of a source package, and the license of the files it contains. For this purpose we create license inclusion graphs. Our results show that more modern reciprocal licenses such as the General Public License v3 tend to include files of less licenses than its previous versions, and that packages under an Apache License tend to contain only files under the same license.

1 Introduction

One way of reducing software development cost is to reuse components. Today Free and Open Source Software (FOSS) has become a common and viable source of components that are ready to be reused. It is possible to find many components from open source project hosted by open source project hosting site such as SourceForge.net[1], Google Code[2] and GitHub[3]. In addition, users can use software component retrieval system such as SPARS-J[1] and Oholo code[4].

The relationship between licenses is complex. It is not trivial to understand when a file with one license can be reused inside a package of another license. Some authors of licenses provide guidelines that try to clarify this; for example, the Free Software Foundation tries to clarify the relationship between the General Public License and other licenses [2]. However, there is a lack of empirical evidence that shows the relationship between the license of the package and the license of the files it contains.

---

[1] sourceforge.net
[2] code.google.com
[3] github.com
[4] ohloh.net

In this paper we describe an empirical study that investigates how different software licenses are reused as white-box components in the software packages found in Fedora, a popular Linux distribution. We show the relationships between the licenses of packages, and the licenses of its files. Our goal is to assist developers, license compliance officers and lawyers in understanding how licenses are actually used.

The contributions of this paper are:

- An empirical study of how the relationship between the license of a package and the license of the files it contains (white-box reuse).
- The definition of *License Inclusion Graphs* that show the licenses that are used inside a packages of a given license.
- The license inclusion graphs from Fedora version 19 for the most popular licenses.

2 Background

The monetary cost of reusing a FOSS component can be zero, but it requires the user to read and accept its FOSS license. In general a software license is a set of permissions that the intellectual property owner grants to the user of the software after a set of conditions have been satisfied (these conditions could be, for example, the payment of a fee). A FOSS license is a software license where the permissions granted include the right to make derivative works of the software and further redistribute those works in exchange to the acceptance of certain conditions (see [3] for a detail description of FOSS licenses, and [4] for a formalization of grants and conditions). For example, the original X11/MIT license grants permissions to "deal in the Software without restriction, including without limitation the rights to use, copy, modify, merge, publish, distribute, sublicense, and/or sell copies of the Software, and to permit persons to whom the Software is furnished to do so"; the only condition it places to grant those rights is "The above copyright notice and this permission notice shall be included in all copies or substantial portions of the Software."

When developers built software by reusing FOSS, either by linking to them or by copying their source code, they have to satisfy the conditions of each of their licenses [4]. Fig. 1 illustrates this problem. In this scenario, the main source code of the project is developed under license A. It reuses two libraries by linking to them (reused as components–i.e. black-box reuse) each under license B and C. To complicate things further source code files from another component (under License D) has been copied into the project. The problem becomes, what license can be used to release the new software system that is compatible with the license of the source code (A) the copied source code (D) and the linked components (B and C). This is not always an easy to answer question, especially when the newly created project reuses many components. For example, Scacchi et. al. [5] examine Google Chrome and showed that it uses 27 components and libraries under 14 different licenses.

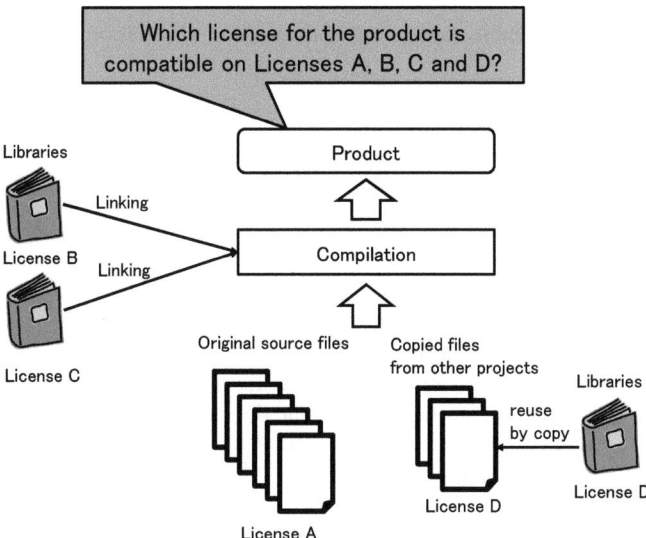

Fig. 1. Example of the challenges of reuse of various licenses. In this example, the code has been developed under license C, but some of the source code was copied from a component with license D. Later this code was compiled and linked with components under license A and B. Can the resulting product be licensed under license A, B, C, D or other license?.

Another problem is the proliferation of licenses. The Open Source Initiative[5], the body responsible for the definition of Open Source, has approved 69 licenses as open source [6] and BlackDuck claims that the Black Duck Knowledge Base includes data related to over 2200 licenses[7]

Previous research has studied license incompatibility. Alspaugh et. al. proposed a model of licensing terms represented by a tuple ⟨actor, modality, action, object, license⟩ and an approach to calculate conflicts of rights and obligations that they implemented on ArchStudio4, a software traceability tool. German et. al. shows integration pattern to solve conflicts of license[4]. However, solving li cense conflicts is not an easy task since it requires legal background. Also, most of the literature regarding license compatibility has focused on the Free Software Foundation licenses (the different versions of the General Public License —GPL— and the Lesser General Public License –LGPL).

In general, files are expected to contain licensing information in them, embedded in the comments of the file. This licensing information describes one or more licenses under which the file is "open-sourced". For this paper we analyzed as a corpus of FOSS the source packages of the Fedora Core 19 distribution. A FOSS system is usually represented in Fedora as a "source" package. A source package is a collection of files, including source code and documentation from where binaries are created.

---

[5] opensource.org

Table 1. Names of common open source licenses and their abbreviations as used in this article

Abbrev.	Name
Apache	Apache Public License
BSD4	Original BSD, also known as BSD with 4 clauses
BSD3	BSD4 minus advertisement clause
BSD2	BSD3 minus endorsement clause
CPL	Common Public License
CDDLic	Common Development and Distribution License
EPL	Eclipse Public License
GPL	General Public License
LibraryGPL	Library General Public License (also known as LGPL)
LesserGPL	Lesser General Public License (successor of the Library GPL, also known as LGPL)
X11/MIT	Original license of X11 released by the MIT

Fedora documents the license of every source package. We call this the *declared license* of the package. This information is intended to inform the user of the package of the legal obligations acquired when using it. This information is usually created manually by inspecting the documentation of the software [8]. Fedora restricts what licenses packages can have in order to be included in the distribution, but this list is fairly comprehensive, currently including 253 different FOSS licenses[9].

Table 1 shows the most common license names and the abbreviation used in this paper. This paper uses a suffix v<number> as means to specify a version of that license; for example, *GPLv2* means *GPL version 2*. If the license name is followed by "+" then the file can be used under such license or newer versions of it; for example *GPLv2+* means the file can be used under the terms of the *GPLv2* or the *GPLv3* (including future versions of the license).

3 License Inclusion

Any FOSS package is licensed under at least one FOSS license, and each of its source code files is expected to be licensed under a FOSS license too. Ideally every file should explicitly indicate its license, although it is not uncommon to find files without a license.

If a software package is of a given license, it does not mean that all the files that compose it are also of the same license. For example, it is widely accepted that files under the MIT/X11 license is included in software that is licensed under any other license, as long as the minimal requirements of MIT/X11 license are satisfied (see above). We call this relationship the **inclusion** of one license into another license. In this case, it means that the MIT/X11 license is included in packages under the GPLv2. Using the inclusion relationship, we can compute the **licensing inclusion graph** of a given collection of software packages as follows:

- Create a node for each of the licenses of packages.
- Create a node for each license found in files.
- For every file f in a package p, we create a directed edge $license(f) \rightarrow license(p)$.

We then define the **license inclusion graph of a package license** as the subgraph that ends in a given package license. All the edges in the graph which share a single destination node and come from different nodes with the same license are merged into a single edge. We put the number of the merged edges as the *weight* of this new edge.

For example, assume that a package in GPLv3 includes 10 source files in BSD2. This relation is represented as a graph including two nodes "BSD2" and "GPLv3" and a single edge from "BSD2" to "GPLv3" with weight 10.

4 Empirical Study

We conducted an case study of FOSS software packages. Its goal is to answer the question *What are the inclusion relationships between licenses of packages and licenses of source code?*

4.1 Subject

For this study we used the 2484 source packages in Fedora Core 19[6]. Each package includes an archive of source files and one or more spec files. The spec files provide metadata of the source package including its license.

4.2 Methodology

To answer the research question we created the license inclusion graphs of package licenses found in Fedora, using the following procedure: First we extracted for each package its declared license, and the license of its files as follows:

1. We extracted its declared license from its spec file.
2. For each source code file in the package, we identified its license(s). For this step we used Ninka[7], a license identification tool with a reported accuracy of 93% [10]. Only 2,013 packages have at least one source code file. The median number of files per package is 60 files, average 748.5, and maximum 125,400.

Ninka is not able to identify the license of all files. Frequently this is because Ninka does not know the way a specific project licenses its files. In that case, Ninka reports "Unknown". In 62.2% of the packages, at least one file was "Unknown" (i.e. in 37.8% of the packages, Ninka identified the licenses of all the files). The distribution of the proportion of "Unknown" files had median of 2.7%, with a 3rd quartile of 25%. This meant that we had incomplete licensing

---
[6] ftp:///ftp.iij.ad.jp/pub/linux/fedora/releases/19/Fedora/source/SRPMS/
[7] ninka.turingmachine.org

information of packages with high ratio of "Unknown". For this reason we decided to remove from the study packages that had a proportion of 50% or more "Unknown" files (328 packages–16% of the total).

Some packages contain more than one spec file. If the licenses of the spec files of a package were different, that meant that some files in the package were distributed under one license, and some under another. For that reason we removed packages with spec files with different licenses. This resulted in another 210 packages being removed from our study.

In total, we kept for our analysis 1,475 packages with at least one source code file, for total of different 511,308 files.

The abbreviation names of licenses in Fedora are different than the names used in Ninka. To make both lists comparable we created an equivalence table. An excerpt of this table is shown in Table 2. The complete table can be found in the replication package (available at http://www.dbms.cs.kumamoto-u.ac.jp/~y-manabe/replication_package/index.html/). Table 3 shows the most frequently found declared licenses in packages. Table 4 shows the number most frequently licenses in source files.

With this information we created the license inclusion graphs of the distribution, and from it the license inclusion graph of the most common licenses.

Table 2. Excerpts from conversion table of license names between Fedora and Ninka

Declared license of package (*Fedora Name*)	License in file (*Ninka Name*)
ApacheSoftwareLicense	Apachev2
ASLv2	Apachev2
BSDwithadvertising	BSD4
EPL	EPLv1
LGPLv2	LibraryGPLv2
LGPL2.1	LesserGPLv2.1
PHP	phpLicV3.01

4.3 Results

Due to space limitations we only show the subgraphs of the most common licenses. The rest are available in the replication package. Figure 2 shows the file-to-package licensing subgraphs for the different versions of the General Public License. The number of package in each source file license node means how many packages have files under the license. The sum of the number of packages in source file is often larger than that in package license node because many packages include license not under same license but under various license. As it can be observed, there is no apparent inconsistency: all licenses of the files (the left node) can be relicensed as the license on the packages (the right node). What is apparent is that the GPLv3+ reuses fewer licenses.

Table 3. Number of packages under each license more than 10 packages)

License name	Packages
GPLv2+	338
LGPLv2+	205
X11mit	154
GPL+ or Artistic	109
BSD	91
GPLv3+	63
ASL 2.0	50
GPLv2+ and GFDL	48
GPLv2	44
GPLv2+ and LGPLv2+	30
LGPLv2	28
LGPLv2+ and GPLv2+	19
LGPLv3+	16
EPL	16

Table 4. Most common licenses used by source files

License Name	Src Files
NONE	111311
EPLv1	70004
GPLv2+	48063
Unknown	32874
LibraryGPLv2+	32745
GPLv3+	29041
BSD3	25570
Apachev2	22099
LesserGPLv2.1+	21733
BSD2	19844
X11mit	15365
GPLv2-classPathExcep	14509
GPLv2	13958
CPLv1	12060

Figure 3 shows the subgraph for the most commonly found permissive licenses. The LesserGPLv2+ shows two main inconsistencies: the use of the GPLv2+ and the GPLv3+. We inspected the files that created this inconsistency, and in most cases, they were in directories that contained the name "test" or "demo", suggesting they were testing and sample programs (most of them very small). It is interesting that these packages would be under the LesserGPL, but the test and demo files under the GPL.

Fedora does not distinguish between the BSD2 and the BSD3, labelling both BSD, as shown in the diagram. In both the BSD and the X11/MIT, the proportion of files without a license is higher than other licenses (> 30%). The "GPL+ or Artistic" is a license that is used mainly for Perl and its modules. It is interesting that most of the files in this graph have no license, or the license "License Same as Perl". This is likely because the template for the creation of a module in Perl uses this license. Notice that in this license, as well as the two Apache, no other license is used. These communities (Perl and Apache) do not seem to reuse code under other licenses. Packages under Apache have the simplest graphs: most files (> 90%) contain the given license, and in the case of Apachev2, only 1.5% of files are under another license. For Apachev1.1 all its files are under the same license.

5 Limitations and Threats to Validity

With respect to the threat of internal validity, in this paper we didn't consider how source files were used. We focused on only relations between a source package and source files included in it. However, not all source files in a package are used in building software. Therefore, we may extract the relations between packages and unused source files. We believe this effect is small.

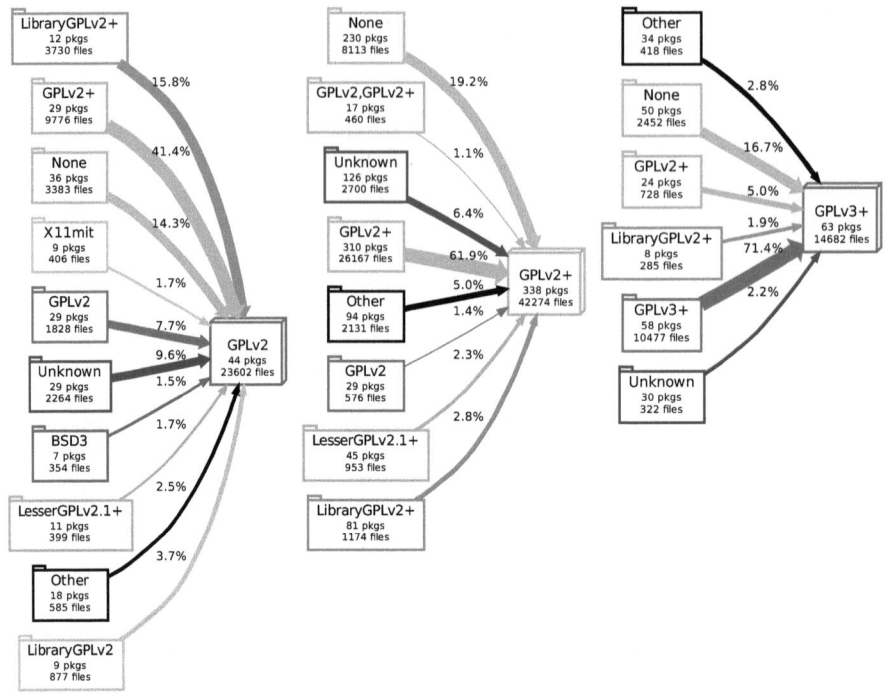

Fig. 2. License inclusion subgraphs for different versions of the GPL

Regarding construction validity, we used Ninka to identify the license of each source file. The quality of the data extracted depends heavily on Ninka's license detection quality. In our previous work [10] we reported that Ninka is accurate in 93% of files. We believe this is enough to represent the most common relationships between licenses. To identify the license of software packages we used the spec files created by the Fedora Project. Our study depends on the accuracy of this information. German et. al. [11] observed that this data is manually generated and an mostly correct. There were very few cases, however, where a package had upgraded to a new license, but the license in the spec file was not updated.

Regarding external validity, we only used source packages in Fedora19. Our results are affected by the selection bias of the Fedora Core Project. To make sure that licenses are comprehensively represented it is necessary to analyze other repositories of FOSS. We plan to do this for future work.

6 Related Work

German et.al[11]. studied license compatibility in software packages by identifying licenses of source packages and source files in them. They found that in general, identifying the license of a package from its source code is not a trivial

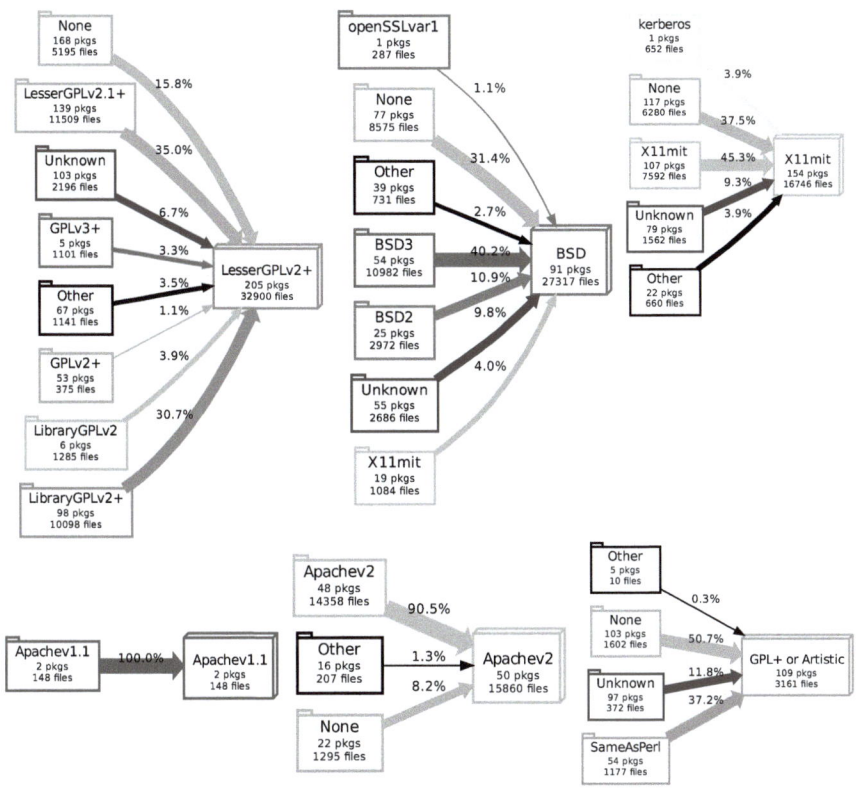

Fig. 3. License inclusion subgraphs of permissive licenses

problem. Our work is an extension to theirs, as we use similar data. The difference is that we focused on the relationships between licenses and not packages.

Stewart et.al [12] addressed the impact of licenses on software projects. Alspaugh et. al [13] have analyzed the requirements that licensing imposes on software. Scacchi et. al [5] have looked into the impact of licenses in the evolution of FOSS software.

7 Conclusions

In this paper we extracted the relationship between the licenses of packages and the licenses of the files are composed of in the Fedora Core 19 distribution. We visualize this information using license inclusion graphs. These graphs show that the different variants of the General Public License are more likely to include other licenses, while licenses such as the Apache tend to contain files only under the same license and they are better at stating the license of their files.

As future work, we would like to analyze the build systems of packages to determine which files are actually part of the binaries. We would like to also explore the licensing relationships between packages, and repeat this study in other collections of FOSS.

Acknowledgements. This work is supported by Japan Society for the Promotion of Science, Grant-in-Aid for Young Scientists (B) (No.24700029), Grant-in-Aid for Scientic Research (S) Collecting, Analyzing, and Evaluating Software Assets for Effective Reuse (No.25220003), and by an Osaka University International Collaboration Research Grant.

References

1. Inoue, K., Yokomori, R., Yamamoto, T., Matsushita, M., Kusumoto, S.: Ranking significance of software components based on use relations. IEEE Trans. Softw. Eng. 31, 213–225 (2005)
2. Foundation, F.S.: Various licenses and comments about them, http://www.gnu.org/licenses/license-list.en.html
3. Rosen, L.: Open Source Licensing: Software Freedom and Intellectual Property Law. Prentice Hall (2004)
4. German, D.M., Hassan, A.E.: License integration patterns: Addressing license mismatches in component-based development. In: Proc. ICSE 2009, pp. 188–198 (2009)
5. Scacchi, W., Alspaugh, T.A.: Understanding the role of licenses and evolution in open architecture software ecosystems. Journal of Systems and Software 85(7), 1479–1494 (2012)
6. Open Source Initiative: Open source licenses, http://opensource.org/licenses/index.html
7. Black Duck Software: Black duck knowledge base, http://www.blackducksoftware.com/products/knowledgebase
8. Callaway, T.: Fedora: Licensing guidelines (2011), https://fedoraproject.org/wiki/Packaging:LicensingGuidelines?rd=Packaging/LicensingGuidelines
9. Callaway, T.S.: Fedora: Software licenses (2013), https://fedoraproject.org/wiki/Licensing:Main?rd=Licensing#SoftwareLicenses
10. German, D.M., Manabe, Y., Inoue, K.: A sentence-matching method for automatic license identification of source code files. In: Proc. ASE 2010, pp. 437–446 (2010)
11. German, D.M., Di Penta, M., Davies, J.: Understanding and auditing the licensing of open source software distributions. In: Proc. ICPC 2010, pp. 84–93 (2010)
12. Stewart, K.J., Ammeter, A.P., Maruping, L.M.: Impacts of license choice and organizational sponsorship on user interest and development activity in open source software projects. Info. Sys. Research 17, 126–144 (2006)
13. Alspaugh, T., Asuncion, H., Scacchi, W.: Intellectual property rights requirements for heterogeneously-licensed systems. In: Proc. RE 2009, pp. 24–33 (September 2009)

Adapting SCRUM to the Italian Army: Methods and (Open) Tools

Franco Raffaele Cotugno and Angelo Messina

Stato Maggiore dell'Esercito Italiano, Italy

Abstract. Many software-related technologies, including software development methodologies, quality models, etc. have been developed due to the huge software needs of the Department of Defense (DoD) of the United States. Therefore, it is not surprising that the DoD is promoting open source software and agile approaches into the development processes of the defense contractors[1]. The quality of many open source product has been demonstrated to be comparable to the close source ones and in many cases even higher and the effectiveness of agile approaches has been demonstrated in many industrial settings. Moreover, the availability of the source code makes open source products attractive for obvious reasons (e.g., security, long term maintenance, etc.). Following this trend, also the Italian Army has started using open source software and promotes its usage into the development processes of its contractors, also promoting agile approaches in many contexts focusing on the SCRUM methodology. This paper provides an overview of the SCRUM development process adopted by the Italian Army for the development of software systems using open source technologies.

1 Introduction

Software systems are becoming larger and larger requiring an increasing amount of resources in terms of effort and budget. In particular, in the military environment where the reliability of systems is of paramount importance, software costs and the length of the development cycles represent rising challenges. To reduce costs and development cycles, Open Source Software (OSS) and Agile Methods (AMs) are opportunities that can be investigated even is such environment.

Considering the development process, AMs have been demonstrated their ability in delivering quality software on time focusing on the value from the point of view of the customer [7, 8, 9]. For this reason, in many cases, AMs are able to satisfy the customer needs and create a strong and trusted relationship with the development team [15]. In particular, SCURM have been selected in many contexts for its focus on agile project management [16].

Traditionally, OSS is characterized by informal development processes, unstable teams, and different levels of quality [10, 11, 12], therefore adopting OSS may be considered risky. However, in many cases, OSS development is supported by software

---

[1] DoD Open Source Software (OSS) FAQ available at:
http://dodcio.defense.gov/OpenSourceSoftwareFAQ.aspx

companies (both large and small) and OSS projects are no more interesting only to enthusiasts and volunteers [6]. These aspects increase the interest in a business use of OSS.

Software companies often use different popular models such as the Capability Maturity Model (CMMI) [1] to assess the level of quality of their development process. However, OSS projects can hardly adopt CMMI, because it is too complex, requires extensive efforts to be used, and does not take into account the characteristics of OSS. For these reasons simpler and more suitable models appeared in the last few years such as the QualiPSo OMM [14], QualiPSo MOST [18], OSMM [2, 3], QSOS [4], OpenBRR [5], etc. Such models focus on different aspects of the production: QualiPSo OMM deals with the quality of the production process, while all the others focus mainly on the quality of the final product. Such models can be used to evaluate OSS and decide if such products can be used in different contexts.

The rest of the paper is organized as follows: Section 2 presents some related work; Section 3 discusses how AMs and OSS can be used in the military environment; finally, Section 4 draws the conclusions and presents future work.

2 Related Work

AMs and SCRUM, in particular, have been demonstrated to be very successful in defense projects [17] helping the military to manage in a new way complex projects that need to adapt quickly to continuously changing environments. AMs are used to address such problems allowing developers to ship working versions of the needed systems reducing the costs and delivering quickly.

During the last 10 years, several open source assessment methodologies have been proposed. Most of them focus mainly on the assessment of the product and do not offer enough metrics to obtain a detailed assessment of the quality of the development process [13].

The first proposed assessment methodology is the Open Source Maturity Model (OSMM) proposed by Cap Gemini [2]. It is the first attempt to standardize the usually ad-hoc assessment approaches of open source projects. The methodology allows the evaluation of 12 product related characteristics and 15 user related characteristics. This way the authors of the methodology allowed a personalized assessment approach based on specific user needs of the FLOSS product.

One year later Navica proposed its own assessment method also called the Open Source Maturity Model (OSMM) [3]. This method is more compact than the one proposed by Cap Gemini. Navica's method is more interesting for our research because it contains some elements that allow a partial assessment of the development process. The method requires the assessment of six project related characteristics. The following four requires some information about the development process: support, documentation, training, and professional services.

In 2004, the Qualification and Selection of Open Source Software (QSOS) [4] method was proposed by a group of OSS developers, users, and enthusiasts. This method has been released with an open source license, allowing an open adoption without any restrictions. The methodology allows an iterative assessment approach. During the first iteration, the number of tools to evaluate are restricted, during the second iteration, their number is additionally limited and so on, until we reach the last stage where we obtain the tool that scores best.

Open Business Readiness Rating (OpenBRR) [5] was specifically designed to allow the assessment of OSS tools that are mature enough to be used by industry. The method was proposed in 2005 and its authors reused many elements introduced in previous OSS assessment methods. The OpenBRR method contains also a few process-oriented elements such as: quality, security, support, documentation, adoption, community, professionalism, and others.

QualiPSo OMM [14] and QualiPSo MOST [18, 19] have been proposed by the QualiPSo consortium in 2009 as two complementary methodologies to assess both the development process (OMM) and the product (MOST). QualiPSo OMM has been designed to be compatible with CMMI allowing companies already using CMMI to implement OMM in an easy way. Moreover, it focuses on different aspects of the development process analyzing it from three points of view: users (consumers of OSS), developers (producers of OSS), and system integrators. Based on such profiles, the methodology provides customized sets of characteristics that require evaluation. QualiPSo MOST has been designed in conjunction to the QualiPSo OMM methodology but focusing on the assessment of the product quality. Most of the assessment criteria defined in MOST can be automated, simplifying the assessment process.

3 Assessing Quality through QualiPSo OMM and MOST

The idea and the structure of OMM and MOST are inspired by the CMMI even if they have been designed specifically to be lightweight and focused on OSS development.

OMM defines three maturity levels (basic, intermediate, and advanced). Each level includes a set of characteristics that need to be assessed. The characteristics are the following:

1. Product Documentation (PDOC)
2. Popularity of the Software Product (REP)
3. Use of Established and Widespread Standards (STD)
4. Availability and Use of a (product) Roadmap (RDMP)
5. Quality of Test Plan (QTP)
6. Relationship between Stakeholders (Users, Developers, etc.) (STK)
7. Licenses (LCS)
8. Technical Environment (Tools, OS, Programming Language, Development Environment) (ENV)
9. Number of Commits and Bug Reports (DFCT)
10. Maintainability and Stability (MST)
11. Contribution to FLOSS Product from SW Companies (CONT)
12. Results of Assessment of the Product by 3rd Party Companies (RASM)

OMM is organized in three levels. Each level includes the characteristics at the lower levels.

The basic level includes:

- Product Documentation (PDOC)
- Use of Established and Widespread Standards (STD)
- Quality of Test Plan (QTP)
- Contribution to FLOSS Product from SW Companies (CONT)
- Licenses (LCS)
- Technical Environment (Tools, OS, Programming Language, Dev Environment.) (ENV)
- Number of Commits and Bug Reports (DFCT)
- Maintainability and Stability (MST)

The intermediate level adds:

- Popularity of the Software Product (REP)
- Availability and Use of a (product) Roadmap (RDMP)
- Relationship between Stakeholders (Users, Developers, etc.) (STK)
- Results of Assessment of the Product by 3rd Party Companies (RASM)

The advanced level includes all the characteristics identified in the basic and intermediate level and adds deepness requiring more details for each characteristic.

The model has been designed to be used in different industrial context from the points of view of the users, the integrators, and the developers. Therefore, it can be used in almost any situation in which OSS is needed [14].

As a complementary methodology, MOST provides the tools to access the quality of the code of OSS. MOST have been designed with the same goals and methodologies of OMM and defines a specific set of characteristics for the evaluation of open source code. [TODO]

4 AMs and OSS in the Military Environment

AMs are designed to support changing requirements through the implementation of an incremental development approach and the close interaction with the customer. In these approaches, the customer becomes part of the development team allowing a better understanding of the problem and a faster development of a system that is able to satisfy user requirements. In the military environment, software systems are very complex and applying AMs is a challenge. Among the available AMs, SCRUM seems to fit better the environment due to its focus on project management with changing requirements.

SCRUM divides the project activities into short iterations named *sprints* that are able to produce a deployable (even if partial) system able to be fully tested by the customer to provide fast feedback and fix the potential problems or requirements misunderstandings.

A SCRUM team includes several actors: a Product Owner (PO), a SCRUM Master (SM), and a Development Team (DT). To adopt SCRUM to develop software systems through a close collaboration between the Italian Army and external contractors, the SCRUM team should be organized in the following way:

- **Product Owner:** he provides the vision of the final product to the team. He guides the development defining the requirements of the product and the related priorities. Therefore, he is the main responsible for the success or failure of the project. His availability to answer questions from the DT and the SM has a significant impact on the development since he is the keeper of the knowledge about the final product. For these reasons, he needs to be a member of the Army with a wide knowledge about the system under development and a clear vision on the expected results. Moreover, he should be able to involve in the testing of the system the people that will use the final system in real operations. Due to the complexity of the role and the wide knowledge needed, the PO is supported by a specific Support Team that includes people from the Army and people from the contractors with deep knowledge about the systems under development and the adopted technologies. The composition of such Support Team will change based on the specific challenges of each sprint. The main role of this team is to help the PO in the definition of the functional and non-functional requirements (including security issues, compliance to standards, usability, performance, etc.). The contribution of the Support Team is very important in particular at the beginning of the project during the definition of the Product Backlog and during the re-prioritization activities required in case of relevant changes in the Backlog. Moreover, in specific conditions related to the technical complexity of a sprint, the Army personnel of the Support Team could take the role of PO just for the duration of the specific sprint. In this way, the PO will be able to guide the development of a large and complex system that can be hardly managed by a single person and require a wide set of skills and knowledge at different levels of granularity. The PO participates actively in the team in the development and in promoting team building activities in conjunction with the SM.

- **SCRUM Master:** he is an expert in the usage of the SCRUM methodology and has a central role in the SCRUM Team coaching the DT and helping in the management of the Product Backlog. The SM helps the DT to address problems in the usage and adaptation of the SCRUM methodology to the specific context in which the team works. Moreover, he leads the continuous improvement that is expected from any agile team. An additional duty of the SM is to protect the DT from any external interferences and lead the team in the removal of obstacles that may prevent the team to become more productive. The SM is selected among the personnel of the IT department of the Army. The SM and the PO have to collaborate closely and continuously to clarify requirements and get feedback. In the Army implementation of SCRUM, the SM has a stronger role focusing on supporting the team in focusing on the requirements and priorities defined by the PO.

- **Development Team:** as any SCRUM development team, people belonging to the team are together responsible for the development of the entire product and there are no specialists focusing on limited areas such as design, testing, development, etc. All the members of the team contribute to the entire production. The DT is self-organized and decides without any external constraints how to organize and execute the Sprint Backlog (the selected part of the Product Backlog that has been identified to be executed in a sprint) based on the priorities defined by the PO on the Product Backlog. The DT usually includes between 4 and 10 people, all expert software developers. In this case, the team includes people form the Army and the contractors, even if the contribution of the contractors is predominant. In the pilot project, the team includes 4 people from a contractor and 2 people from the Army, in subsequent projects we expect to be able to replicate such unit to manage more sprints in parallel.

The SCRUM methodology is not a one-size-fits-all approach in which all the details of the process are pre-defined and the development team have to stick with it without any modification. On the contrary, SCRUM defines a high-level approach and a state of mind for the team promoting change management and flexibility in the work organization aimed at satisfying the customer needs. As all the other AMs, SCRUM defines values, principles, and practices focused on close collaboration, knowledge sharing, fast feedback, tasks automation, etc.

The SCRUM development process starts from a vision of the PO. Such vision is a wide and high-level definition of the problem to address that will be refined and narrowed during the development through the Backlog Grooming. The Backlog Grooming is an activity that takes place during the entire development process focusing on sharpening the problem definition, pruning redundant and/or obsolete requirements, and prioritizing the Backlog. Moreover, in this phase, the PO defines the scenarios and the criteria used to test the (and accept) the user stories.

Ones defined the Product Backlog, during the Planning Meeting, the PO and the customers/stakeholders define the detailed user stories assigning them a priority that are used to organize the single sprints. During the sprint planning phase, the PO and the DT agree on the objectives of the sprint. The DT decompose features in the Backlog into detailed tasks. Such list of tasks becomes the Sprint Backlog that will be implemented during the sprint execution producing the (incremental) shippable product. At the end of each sprint, a sprint review takes place to verify the progress of the project comparing it to the expectations. The results of the review are used to adapt the Product Backlog modifying, removing, and adding requirements. This activity is very important for the success of a project and involves all the member of the SCRUM Team and the interested stakeholders. Additionally, other SCRUM Teams could join the meeting if the project requires the collaboration of more teams. This activity is very important to align the teams with the stakeholders and to manage the project properly even if changes happen. The Product Backlog is dynamic, changing continuously collecting all the requirements, issues, ideas, etc. and should not be considered in a static way, it is designed to support the variability of the environment.

After the Sprint Review, there is the Sprint Retrospective in which the PO, the SM, and the DT analyze together the sprint just finished to evaluate its effectiveness and to identify problems and opportunities to improve.

In the specific context of the Army, the SCRUM Team should be organized in the following way:

1. The PO defines the Product Backlog that, based on the specific objective of the sprint, may include:
    a. High-level views (e.g., Operational Views) describing the functional properties from the operational point of view that can be easily shared and discussed with the operational units.
    b. A natural language description of the user stories of the items of the Product Backlog
2. Based on the Backlog, the PO identify the proper people to include in the Support Team. At this level, only candidate people are identified but no actual assignment is performed because the assignment is performed at each sprint based on the specific competences needed.
3. Once defined the Backlog, the DT defines the Sprint Backlog and the PO assign specific people to the Support Team selecting them from the pool identified at the beginning and satisfying the needs of the current sprint.
4. After these activities, the DT can start the execution of the sprint. Once the development activities are completed (including all the functional and non-functional testing), the current design documents can be generated automatically through tools for code reverse engineering. Such documentation will be used at the beginning of the subsequent sprint.
5. At the end of the execution, a review is performed involving the entire SCRUM team and the stakeholders (mainly from the operational units) able to assess the usability of the system in the real context.
6. At the end of the sprint, a retrospective takes place involving the PO, the SM, and the DT. Such activity focuses on improving the effectiveness of the team through process improvement strategies and evaluating new approaches.

In the context of the Army, testing has a very important role due to several constraints:

- **Testing at development level:** testing activities performed by the development team in the development environment and in a specific testing environment with the same characteristics of the deployment environment. Such activities are the traditional ones performed in any Agile team.
- **Testing at operating unit level:** testing activities performed by the operating units in limited (but real) environments to verify the actual effectiveness of the developed systems in training and real operating environments.
- **Testing at certification level:** testing activities performed to verify the compliance of the developed systems to the national and international (NATO) standards that are needed for authorizing the usage of the systems in national and joint NATO operations.

5 Conclusions and Future Work

In this paper, we have presented an overview of the usage of QualiPSo OMM and MOST for the evaluation of the quality of OSS and SCRUM for addressing the development needs of the Italian Army. This is just an initial step for providing a comprehensive analysis on how OSS and AMs can be used in such environment reducing costs and increasing the effectiveness of the development teams.

References

1. Amoroso, E., Watson, J., Marietta, M., Weiss, J.: A process-oriented methodology for assessing and improving software trustworthiness. In: 2nd ACM Conference on Computer and Communications Security (1994)
2. Duijnhouwer, F.-W., Widdows, C.: Capgemini Expert Letter Open Source Maturity Model, Capgemini (2003)
3. Goldman, R., Gabriel, R.P.: Innovation Happens Elsewhere - Open Source as Business Strategy. Morgan Kaufmann, Boston (2005)
4. Atos Origin, Method for Qualification and Selection of Open Source Software (QSOS), http://www.qsos.org
5. Wasserman, A., Pal, M., Chan, C.: Business Readiness Rating Project, BRR Whitepaper (2005), http://www.openbrr.org/wiki/images/d/da/BRR_whitepaper_2005RFC1.pdf
6. Dueñas, C.J., Parada, H.A., Cuadrado, G.F., Santillán, M., Ruiz, J.L.: Apache and Eclipse: Comparing Open Source Project Incubators. IEEE Software 24(6) (November/December 2007)
7. Sillitti, A., Succi, G.: Requirements Engineering for Agile Methods. In: Aurum, A., Wohlin, C. (eds.) Engineering and Managing Software Requirements. Springer (2005)
8. Janes, A., Remencius, T., Sillitti, A., Succi, G.: Managing Changes in Requirements: an Empirical Investigation. Journal of Software: Evolution and Process 25(12), 1273–1283 (2013)
9. Di Bella, E., Fronza, I., Phaphoom, N., Sillitti, A., Succi, G., Vlasenko, J.: Pair Programming and Software Defects – a large, industrial case study. IEEE Transaction on Software Engineering 39(7), 930–953 (2013)
10. Di Bella, E., Sillitti, A., Succi, G.: A multivariate classification of open source developers. Information Sciences 221, 72–83 (2013)
11. Scotto, M., Sillitti, A., Succi, G.: Open Source Development Process: a Review, International Journal of Software Engineering and Knowledge Engineering. World Scientific 17(2), 231–248 (2007)
12. Jermakovics, A., Sillitti, A., Succi, G.: Exploring collaboration networks in open-source projects. In: Petrinja, E., Succi, G., El Ioini, N., Sillitti, A. (eds.) OSS 2013. IFIP AICT, vol. 404, pp. 97–108. Springer, Heidelberg (2013)
13. Petrinja, E., Sillitti, A., Succi, G.: Comparing OpenBRR, QSOS, and OMM Assessment Models. In: Ågerfalk, P., Boldyreff, C., González-Barahona, J.M., Madey, G.R., Noll, J. (eds.) OSS 2010. IFIP AICT, vol. 319, pp. 224–238. Springer, Heidelberg (2010)
14. Petrinja, E., Nambakam, R., Sillitti, A.: Introducing the Open Maturity Model. In: 2nd Emerging Trends in FLOSS Research and Development Workshop at ICSE 2009, Vancouver, BC, Canada, May 18 (2009)

15. Sillitti, A., Ceschi, M., Russo, B., Succi, G.: Managing Uncertainty in Requirements: a Survey in Plan-Based and Agile Companies. In: 11th IEEE International Software Metrics Symposium (METRICS 2005), Como, Italy, September 19-22 (2005)
16. Schwaber, K.: Agile Project Management with Scrum. Microsoft Press (2004)
17. Crowe, P., Cloutier, R.: Evolutionary Capabilities Developed and Fielded in Nine Months. CrossTalk: The Journal of Defense Software Engineering 22(4), 15–17 (2009)
18. del Bianco, V., Lavazza, L., Morasca, S., Taibi, D., Tosi, D.: An investigation of the users' perception of OSS quality. In: Ågerfalk, P., Boldyreff, C., González-Barahona, J.M., Madey, G.R., Noll, J. (eds.) OSS 2010. IFIP AICT, vol. 319, pp. 15–28. Springer, Heidelberg (2010)
19. Lavazza, L., Morasca, S., Taibi, D., Tosi, D.: An empirical investigation of perceived reliability of open source java programs. In: 27th Symposium on Applied Computing (SAC, Riva del Garda, Italy (2012)

Applying the Submission Multiple Tier (SMT) Matrix to Detect Impact on Developer Interest on Open Source Project Survivability

Bee Bee Chua

University of Technology, Sydney

Abstract. There is a significant relationship between project activity and developer interest on Open Source (OS) projects. Total project activity submission count number can be an indicator for gauging developer interest. The higher the project activity submission of a project is, the larger developer interest in a project. My paper proposed that applying a Submission Multiple Tier (SMT) matrix can detect the impact of developer interest on project activity. Results showed more volume of OS projects with low project activity than high. Activity submission results also showed that developers are more likely to review than correct projects, with the first priority to find and fix bugs. Further research is needed to determine the impact of project activity type on developer motivation to contribute, participate and support OS projects.

1 Introduction

For any Open Source (OS) software to survive, appropriate OS infrastructure must be in place. These include the viability of project activities increases, high quality codes must produce, quality and qualifying peer developer networks expandable and OS models adoption are strategies for building an effective infrastructure OS for projects.

Developer support and collaboration is vitally important for maintaining an OS project's survivability, especially if the project is new. Project activity is crucial to survival. Making project activity viable is important as OS coordinators and sponsors can monitor project activity performance, including growth, reputation and status.

This paper introduces the Submission Multiple Tier (SMT) matrix, which is based on a structured, hierachical tier for detecting submission pattern similarity over multiple project activities. The aim of the SMT is to allow possible detection of developer interest by activity type and project population. The SMT can assist OS coordinators and sponsors by: detecting projects with low survival difficulty based on the low submission counts from project activities; and identifying specific project activity interests of OS developers by the high submission counts.

The SMT was constructed to provide a summary of a total project number on each project acitvity by submission tier, so that OS spon sors and resource planning coordinators can provide better OS infrastructure support (for instance, developing a network strategy to increase developer interest on a particular project activity with extremely high number of unfixed bugs.

This paper is divided into five sections: section 2 discusses existing literature on OS infrastructure to support survivability, particularly project activity and developer interest; section 3 introduces the SMT and outlines the procedures for applying the SMT on a Sourceforge.net dataset; section 4 presents results; and section 5 conclusions and future work.

2 Open Source Infrastructure: Project Activity and Developer Interest

A number of studies have investigated OS variables for survival analysis, focusing on programming languages, licences, developer interest, operating systems and end user interest [1-10]. Studies [4-11] also used these variables to measure other project performance outcomes, such as successability, popularity, efficiency and effectiveness. They are essential variables for supporting OS infrastructure.

Good infrastructure is essential for OS project survival. The term 'infrastructure' is defined as "an underlying base or foundation especially for an organization or system" [11]. In this paper, infrastructure is classified as basic or advanced. Basic infrastructure is essential components needed to develop and support OS software (OSS), such as programming languages, licences, operating systems and developers, whereas advanced infrastructure refers to activities positively or negative influencing basic components of OS programs, projects and products. An example of an advanced OS infrastructure is project activity status, which influences developer interest and impacts project download. In addition to a high OSS adoption rate, project activity and high developer interest are vital infrastructure components to enable project survival.

Current OS literature discusses how project activity can be used to predict project success and popularity [3,4,9,10]; however, there is minimal work analysing this variable from a survivability perspective. My research showed non-surviving projects were physically removed from an OS server, irrespective of project activity. This was in contrast to the study by Samoladas [12] that did not emphasise zero project activity for non-surviving projects, instead classifying project activity status as inactive. As a result their analysis on non-surviving projects showed an absence of zero project activity, which is a crucial variable to consider.

3 Methods

The following procedures were followed to analyse project activity submissions and classify projects by survival status.

1. Sourceforge was chosen as it is the world's largest repository of OS projects, with over 100,000 projects and over a million registered users [13,14]. Sourceforge.net data dump files for 2010 and 2012 were downloaded ("stats_project_all"). Each file had 13 standard variables on download: project ID, number of developers, number of submissions opened and closed on bugs, support, patch, artefact and task. However only 10 standard variables were related to project activity and were selected for data review, creating 9 columns for the Multiple Submission Multiple Tier (SMT) Structure (Table 1).

Table 1. Submission Tier Matrix Structure

Variable Type	Activity	Description
Nominal	Bugs opened, support opened, patch opened, artefact opened, task opened	Number of submission opened
Nominal	Bugs closed, support closed, patch closed, artefact closed, task closed	Number of bugs opened
Nominal	Submission Tier 1 to 6	Submission Tier 1 to 6

2. Data were cleaned to remove blank submissions or projects with zero bug submissions opened, leaving 89,002 projects in 2010 and 89,916 in 2012. In 2010, submissions closed totalled 73,016 and for 2012, 66,777. Each project had up to five project activities (bug, support, patch, artefact and task). The project activity submission ranges varied.

3. The projects were then classified as surviving or non-surviving using the SMT. Project ID was used as the key identifier at both time points. If the same project number was present in 2010 and 2012 then it confirmed the project had survived (hosting) and had not been removed from the Sourceforge database. If a 2010 project ID had no corresponding 2012 project ID and zero bug submissions opened (a value marked with '0') then we confirmed the project was deleted, i.e., no longer hosting. By 2010, 435 non-surviving projects had been deleted and archived in Sourceforge. These projects were examined further to determine the relationship between developer size and project activity. Table 2 shows non-surviving projects against project activity. Figure 1 shows effect of activity on developer size of non-surviving projects.

Table 2. Non-surviving Projects Against Project Activity

No. of Developers	Projects with Activity Submissions *	Projects With No Activity Submissions *
0	1	0
1	6	411
2	7	6
3	1	1
7	1	0
15	1	0

* Bug, support, artefact, patch, task.

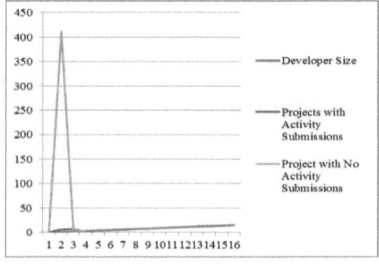

Fig. 1. Developer Size and Project Activity

4. The project total was counted for each activity case for 2010 and 2012, based on their submission number and per SMT tier in section 4.

4 Results

Project Activities for Open Submissions Tiers

Table 3 illustrates the opened submissions pattern of multiple project activities, showing open submission tiers, total project count for each tier, project activities for 2010 and 2012 and project activity performance in terms of change between 2010 and 2012. Each project activity cell shows the project total number. The highest project population was bug submissions opened (26,453 projects in 2010 and 26,636 projects in 2012). Submission tier 1 had the lowest submission count (1–10 submissions), with bugs, support and patch being the three most active project activities in 2010 and 2012: bugs 26,453 (yr10) and 26,636 (yr12); support 17,053 (yr10) and 17,055 (yr12); patch 7402 (yr10) and 7112 (yr12). Projects in this tier could have a high source code quality and, as a result, low active project activity submission by developers for bugs, support, patch, artefact and task

Table 3. SMT Results for Open Submissions

Submission Tier	Bug Submission Opened 2010	Bug Submission Opened 2012	Support Submission Opened 2010	Support Submission Opened 2012	Patch Submission Opened 2010	Patch Submission Opened 2012	Artefact Submission Opened 2010	Artefact Submission Opened 2012	Task Submission Opened 2010	Task Submission Opened 2012
Tier 1 1-10	26453	26636	17053	17055	7402	7112	7140	3049	1133	1335
Tier 2 11-100	6751	6719	5030	4983	4363	4636	4607	6156	1371	1338
Tier3 101-1000	1477	1495	1211	1205	1378	1413	1391	1813	1243	3771
Tier4 1001-10,000	128	132	99	105	167	174	168	344	374	372
Tier5 10,001-100,000	2	2	5	8	9	10	9	15	38	38
Total projects for each project activity	34811	34984	23398	23356	13319	13345	13315	11377	4159	6854

Project Activities Closed Submission Tiers

Table 4 displays six closed submissions tiers, one total project for each tier and ten other project activities for 2010 and 2012. The cells show total projects for each project activity closed submission. Closed bug submissions were the most active (19,258 closed in 2010 and 22,752 in 2012), with 1395 in 2010 and 1665 in 2012 for artefact submissions closed, and 449 and 454 respectively in 2010 and 2012 for task submission closed (see Table 4).

Table 4. SMT on Submissions Closed

Submission Tier	Bug Submission Closed 2010	Bug Submission Closed 2012	Support Submission Closed 2010	Support Submission Closed 2012	Patch Submission Closed 2010	Patch Submission Closed 2012	Artefact Submission Closed 2010	Artefact Submission Closed 2012	Task Submission Closed 2010	Task Submission Closed 2012
Tier 1 1-10	19258	22752	9558	10398	3766	3049	1395	1665	449	454
Tier 2 11-100	6165	6215	3934	4050	3592	6156	2239	2067	750	677
Tier3 101-1000	24476	1723	1158	1161	1646	1813	1680	1603	854	807
Tier4 1001-10,000	171	177	135	138	264	344	432	288	366	1162
Tier5 10,001-100,000,0	1	5	3	5	10	15	18	18	18	35
Tier 6 100,001-100,0000	0	0	0	0	0	0	1	0	1	0
Total projects for each project activity	50071	30872	14788	15752	9278	11377	5765	5641	2438	3135

For tier 1, opened and closed submissions were similar and had the highest project populations. There were more opened than closed tier 1 bug submissions, suggesting developers were more interested in reporting bugs than correcting them. This could be due to developers finding it easier to report than correct during review, as no solution is required. The tier 2 project population ranged from 2000 to 6500 closed submissions: 6165 projects in 2010 and 6215 projects in 2012 for closed bug submissions; 3934 and 4050 in 2010 and 2012 respectively for closed support submissions; 3592 and 6156 for closed patch submissions; 2239 and 2067 for closed artefact submissions; and 750 projects in 2010 and 677 projects in 2012 for closed task submissions.

5 Conclusions and Future Work

For the SMT matrix, three common survival patterns for project activity were found from the two submission patterns: 1) projects survive with a minimum of one project activity submission; 2) the most activity submissions – either open or closed – are bug submissions; and 3) more submissions can positively influence project survival from the support perspective. The submission patterns also revealed that many projects have very low project activity submissions and that developers review more than correct, motivated by project type rather than project activity. We plan to extend our work by investigating project submission based on project activity to confirm survivability to 5, 10 and 15 years. We will also validate the SMT on other OS repositories.

References

1. Hars, A., Ou, S.: Working for free? – motivations of participating in open source projects. In: Proceedings of the 34th Annual Hawaii International Conference on Systems Sciences (2001)
2. Mockus, A., Fielding, R.T., Herbsleb, J.D.: Two case studies of open source software development: Apache and Mozilla. ACM Transactions on Software Engineering and Methodology 11(3), 309–346 (2002)
3. Crowston, K., Howison, J., Annabi, H.: Information systems success in free and open source development: Theory and measures. Software Process and Practice 11(2), 123–148 (2006)
4. Midha, V., Palvia, P.: Factors affecting the success of open source software. The Journal of Systems and Software 85(4), 895–905 (2012)
5. Chen, S.: Determinants of survival of open source software: An empirical study. Academy of Information and Management Sciences Journal 13(2), 119–128 (2010)
6. Lee, H.W., Kim, S.T., Gupta, S.: Measuring open source software success. Journal of Omega 37(2), 426–438 (2009)
7. Choi, N.J., Chengalur-Smith, S.: An exploratory study on the two new trends in open source software: End-users and service. In: Proceedings of the 42nd Hawaii International Conference on System Sciences (2009)
8. Wang, J.: Survival factors for free open source software projects: A multi-stage perspective. European Management Journal 30(1), 352–371 (2012)

9. Subramaniam, C., Sen, R., Nelson, M.L.: Determinants of open source software project success: A longitudinal study. Journal of Decision Support Systems 2(46), 576–585 (2009)
10. Ghapanchi, A.H., Aurum, A.: Competency rallying in electronic markets: Implications for open source project success. Journal of Electronic Markets 22(2), 11–17 (2012)
11. http://en.wikipedia.org/wiki/SourceForge
12. Samoladas, I., Angelis, L., Stamelos, I.: Survival duration on the duration of open source projects. Journal of Software and Information Technology 52(9), 902–922 (2010)
13. Christley, S., Madey, G.: Analysis of Activity in the Open Source Software Development Community. In: Proceedings of the 40th Hawaii International Conference on System Sciences (2007)
14. http://www.sourceforge.net/

FOSS Service Management and Incidences

Susana Sánchez Ortiz and Alfredo Pérez Benitez

University of Informatics Sciences (UCI), Road to San Antonio de los Baños,
Km 2½, Torrens, Havana, Cuba
{ssanchez,apbenitez}@uci.cu

Abstract. The Free Open Source Software (FOSS) solutions have been reaching a high demand, usage and global recognition, not only in the development of applications for companies and institutions also in the management of services and incidents. With the upswing of Information Technology (IT), the development of tools that enable the reporting of problems and incidents on any organization or company is necessary. Every day you need more applications, software generally, that make easier the user's actions. This paper describes the need to use these tools and recount the development of a web application that allows the management of reports and incidents from users of Nova, the GNU/Linux Cuban distribution.

Keywords: FOSS, service management and incidences.

1 Introduction

The services of Information Technology (IT) are more and more complex, which means that their management is most needed to remain efficient. In recent years IT have begun to rely on technological tools and Service Level Agreements(SLAs), which has allowed the development of a more efficient work, streamline their processes and transactions and have information for decision-making in real time.

In order to improve support and management in the area of information have been developed in the world the Systems and Service Management Incidents; technologies that manage incidents and routine requests for new services. These systems provide a set of interrelated components, which owe their existence to the needs of IT services and the human factor that has been increasingly demanding quality management. In Cuba, this subject has not been materialized on a large scale in all enterprises and institutions. The University of Informatics Sciences (UCI) has not been exempt of this situation, this affects their important change process, starring Free Open Source Software (FOSS) solutions. Because the importance of the migration process and the need to maintain with maximum efficiency the functioning of all the girded areas in the process, it has become necessary to have tools to support it.

In Cuba has been developed the GNU/Linux distribution Nova oriented to novice users who are facing a migration process from Microsoft Windows to GNU/Linux. Once the operating system provides its services in the country, it is impossible that their developers give support to all the questions raised by the users, considering that in many cases these concerns can be clarified or solved using new technologies.

In correspondence to the problem presented, the work will be focused on the presentation of the NovaDesk web application, its features and functionalities.

2 Development

2.1 Technologies of Incidences and Services Management

In order to develop a system to manage the incidents that could present a user in its work with Nova was developed an investigation about the management of services and incidents and associated technologies.

The Office of Government Commerce in the United Kingdom [1] defines the management of services as a set of specialized organizational capabilities which provide value to customers in the form of services. The services provide value to customers and facilitate achieve their objectives at less cost and less risk, because the responsibility is assumed by the contracted company [1]. Some of the technologies of services management and incidences are the Service Desk, Help Desk and Call Center. The Call Center can be defined as a place where telephone communications are concentrated in a company [2]. A Help Desk is a part of the technical support group established by an organization to keep their computers running efficiently [3].

The Service Desk have been perceived traditionally as a group of specialists who collect everything and who, hopefully, have the appropriate technical skills to answer practically any questions or complaints. As is represented in Information Technology Infrastructure Library (ITIL) [4], this discipline of Service Desk has evolved to the point that can be executed with a high degree of efficacy [5]. Differs from a Call Center or Help Desk in order that it have a larger scope and is more focused on the customer, because it is responsible to facilitate the integration of business processes on IT infrastructure [5]. One of the most important processes developed by the Service Desk is the management of incidents. For this reason the Service Desk should be supported by Help Desk technology that performs this type of management by excellence.

2.2 Development of NovaDesk

After completed the study of the technologies discussed above and considering that ITIL proposes the use of the Service Desk was decided that this would be the technology to be used. It was then performed an investigation of the open source tools that allow the services management and incidences. It was presented the problem of the low existence of open source applications based on this technology. So the system was initially developed in the Help Desk OneOrZero (v1.8). But the base technology was changed due to several setbacks among which were that the last version of this system (v2.5.1) did not allow it to be used freely.

Due to the amount of skills and to be containers of features that make possible the implementation of the processes and be able to satisfy greatly shortcomings identified in OOZ three options were selected: IRM Information Resource Manager, GLPI and ExoPHPDesk. It was made a comparison and concludes that the most suited to the needs expressed was GLPI.

GLPI is the Information Resource-Manager with an additional Administration-Interface. It has enhanced functions to make the daily life for the administrators easier, like a job-tracking-system with mail-notification and methods to build a database with basic information about your network-topology. GLPI is under the GPL license. The principal functionalities of the application are the precise inventory of all the technical resources and management and the history of the maintenance actions and the bound procedures [6].

GLPI has a group of characteristics [6] among which are: multi-entities and multi-users management; multiple authentication system (local, LDAP, AD) and multiple servers; permissions and profiles system; complex search module; export system in PDF, CSV, SLK, PNG and SVG; management of problems; tracking requests opened using web interface or email; SLA with escalation (customizable by entity); Assignment of interventions demands to the technicians; statistics reports by month, year, total in PNG, SVG or CSV; management of a basic system of knowledge hierarchical. Furthermore, this system has a large number of advantages among which include: reducing costs, optimizing resources, rigorous management of licenses, high quality, satisfying usability and security.

2.3 Functionalities of NovaDesk

Besides the features of GLPI, NovaDesk has a group of features among which are: chat room for technicians, which allows exchanging problems and solutions between them; email notification of monitoring incidents; improved attention through user's chat; edition of chat conversations that provide solutions for uploading to the knowledge base; statistical monitoring of time spent in chat; management reporting assignment; account registration, introducing obligatory and optional fields for the user; password recovery for registered users; user validation; display data from the hardware which users are working through a query of the OCS inventory database and management of the fields of incidence, allowing to adapt to the infrastructure of the organization.

NovaDesk also has an expert system that allows the management of its knowledge base. This system has several functions within which are: manage rules and questions of expert system; manage and identify the category of the problems; fill the knowledge base; identify problems; show answers and solutions of problems; send the solutions of problems by email and in the case in which the expert system does not identify the problem proceed to register as a new incident.

After implemented the expert system has been developed an analysis of their behavior that consisted of verification, validation and evaluation. This process allow to check that the behavior of the expert system is similar to the behavior of a human expert and because of this, the expert system can diagnose the problems presented by users in their work with Nova at the same way that would be performed by a human expert.

2.4 Importance of Services Management and Incidences with NovaDesk

There are a large number of issue tracking system and bug tracking system like Bugzilla, Jira, Redmine, Mantis, WebIssues. When comparing NovaDesk with these tools, it demonstrates that all are at the same level.

The development of NovaDesk has allowed the management of incidents presented by the users to interact with Nova. It has enabled this process more pleasant and accessible to the user, because it can clarify their doubts in different ways, either using the chat, the expert system or recording a new incident. It is also a tool that has been customized for this function and adapts to all companies and Cuban institutions that wish to face a process of migration to free and open source software. NovaDesk also uses the Apport [7] service of the operating system and enables automatic report generation with the information provided, classifying this information on the category of the report and automatically assigning this incident to the group of developers of Nova.

3 Conclusions

NovaDesk provides great benefits to users of Nova, because it enables them to make reports related to deficiencies and disagreements. The purpose of this tool has been to optimize the system functionalities based on the registered incidents and give immediate response to the users concerns. The added features and the change of OOZ by GLPI are the basis for achieving the maximum expression of the processes proposed in ITIL, the implementation of a future Service Desk. The expert system constitutes a way for users to be able to solve the most common problems that can presented without the dependency of a human expert.

References

1. OGC ITIL v3- Estrategia del servicio. 1ª publicación. Reino Unido: TSO (The Stationery Office). ISBN: 978 011 331158 3 (2009)
2. Babylon Team. Definición de Call Center, http://diccionario.babylon.com/
3. mmujica. Tecnología de Información: Help Desk,
   http://mmujica.wordpress.com/2008/10/09/help-desk
4. IT Management Fundamentals - ITIL - What is ITIL? (2013),
   http://itil.osiatis.es/ITIL_course/it_service_management/
   it_management_fundamentals/what_is_itil/what_is_itil.php
5. Panorama IT. Service Desk,
   http://www.panoramait.com/ItileISO20000_ServiceDesk.aspx
6. GLPI - Gestionnaire libre de parc informatique (2013),
   http://www.glpi-project.org/spip.php?article53
7. Apport - Ubuntu Wiki (2014), https://wiki.ubuntu.com/Apport

Open-Source Software Entrepreneurial Business Modelling

Jose Teixeira[1] and Joni Salminen[2]

[1] University of Turku (UTU), Finland
Turku Centre for Computer Science (TUCS), Finland
jose.teixeira@utu.fi
http://www.jteixeira.eu
[2] Turku School of Economics (TSE), Finland
joni.salminen@tse.fi

Abstract. This poster aims to facilitate business planning of bootstrapping entrepreneurs who are developing a high-tech business by open-source approach. It draws on scholarly works on business modelling and open-source software to provide a practical tool for entrepreneurs establishing a business by open-source approach. Built on top of established business modelling frameworks, the Open-Source Software Entrepreneurial Business Modelling (OSS_EBM) can be a useful strategic management and entrepreneurial tool. It enables strategists and entrepreneurs to describe, design, challenge, invent, brainstorm, pivot, analyze and improve upon open-source business models.

1 Introduction

The open-source software community has already established itself as a valuable source of innovation [1–3]. However, how to make money with open-source software remains an interesting question intriguing both researchers and practitioners [4–6]. This paper draws on research on business modelling [7–9] and open-source software [6], [10–14] to provide a practical tool for bootstrapping entrepreneurs developing a high-tech business by open-source approach. Built on top of other previously established business modelling frameworks [8], [9], the Open-Source Software Entrepreneurial Business Modelling (OSS_EBM) is a conceptual tool that aims to be useful in entrepreneurial planning and strategic management of open source ventures. Specifically tailored for the high-tech software business, the OSS_EBM framework enables high-tech strategists and entrepreneurs to describe, design, challenge, invent, brainstorm, pivot, analyze and improve open-source business models. This poster targets mostly practitioners interested in setting up a business with the help of open-source technology. Our earlier work reveals there is a need for better tools of business planning among open source ventures [15].

2 Methods

The method of OSS_EBM is to turn earlier scholarly works to practice in the format of a visual canvas. Exploiting the success of the Business Model Canvas [8], the popular business modelling framework was customized for the specificities of the high-tech software business and complemented with "platform thinking" and ecosystem features from the VISOR framework [9]. Moreover, the fusion of the two previously mentioned business modelling frameworks was complemented with open-source research knowledge on software licensing [12–13] and business models [4–6]. The OSS_EBM framework is thereby a simple multi-disciplinary combination of research in business models and open-source software, turned into a visual canvas in the purpose of guiding practitioners in setting up an open-source based business.

We are currently in this process of testing how the tool can best be applied by entrepreneurs. Earlier versions of the OSS_EBM framework were introduced in an e-commerce university course targeting both master and doctoral students. Moreover, this framework was presented at two events of early-stage startup business incubators. Four workshops and free consultation were conducted with startup teams interested in developing a software business by open-source approach. Feedback was collected informally from bootstrap entrepreneurs, serial entrepreneurs and other personnel of the incubators; leading to some changes on the OSS_EBM framework till date. Modifying the tool further is currently underway.

3 Finding, Implications and Future Work

So far, practitioners' impressions on the relevance of the OSS_EBM framework vary widely. Most recognized the value of this framework, for being specifically tailored for the high-tech software business; however, most users of the OSS_EBM framework still report difficulties in understanding how their business ideas can generate monetary results with open-source software. Even if the framework does not help all its users in developing a business by open-source approach, all users claimed to have learned to a large extent about open-source software after using the framework. We wish then that the OSS_EBM framework will continue developing while helping high-tech entrepreneurs develop, or at least consider developing, businesses by open-source approach. In particular, more thorough action research studies are on the way; guiding and observing focal startups in the use of this tool. Our ultimate goal is to offer technology-minded founders a useful tool for managing the wide selection of business model components, including revenue models, and for crafting the right combination in their particular case.

References

[1] Henkel, J.: Selective revealing in open innovation processes: The case of embedded Linux. Research Policy 35(7), 953–969 (2006)
[2] Lerner, J., Tirole, J.: The open source movement: Key research questions. European Economic Review 45(4), 819–826 (2001)

[3] Weber, S.: The success of open source, vol. 368. Cambridge University Press (2004)
[4] West, J., Gallagher, S.: Challenges of open innovation: the paradox of firm investment in open source software. RD Management 36(3), 319–331 (2006)
[5] Hecker, F.: Setting up shop: the business of Open-Source software. IEEE Software 16(1), 45–51 (1999)
[6] Wasserman, T.: Building a Business on Open Source Software (2009), http://works.bepress.com/tony_wasserman/3/ (accessed November 14, 2013)
[7] Osterwalder, A., Pigneur, Y., Tucci, C.L.: Clarifying business models: Origins, present, and future of the concept. Communications of Association for Information Systems 16(1), 1–25 (2005)
[8] Osterwalder, A., Pigneur, Y.: Business Model Generation: A Handbook for Visionaries, Game Changers, and Challengers. Wiley (2010)
[9] Sawy, O.A.E., Pereira, F.: Business Modelling in the Dynamic Digital Space: An Ecosystem Approach. Springer (2013)
[10] Aksulu, A., Wade, M.: A comprehensive review and synthesis of open source research. Journal of Association for Information Systems 11(11), 576–656 (2010)
[11] Crowston, K., Wei, K., Howison, J., Wiggins, A.: Free/Libre open-source software development: What we know and what we do not know. ACM Computing Surveys CSUR 44(2), 7 (2012)
[12] Lindman, J., Rossi, M., Paajanen, A.: Matching open source software licenses with corresponding business models. IEEE Software 28(4), 31–35 (2011)
[13] Lindman, J.: Not Accidental Revolutionaries: Essays on Open Source Software Production and Organizational Change. Aalto University, School of Economics, Department of Information and Service Economy (2011)
[14] Salminen, J., Teixeira, J.: Fool's Gold? Developer Dilemmas in a Closed Mobile Application Market Platform. In: Järveläinen, J., Li, H., Tuikka, A.-M., Kuusela, T. (eds.) ICEC. LNBIP, vol. 155, pp. 121–132. Springer, Heidelberg (2013)
[15] Teixeira, J., Salminen, J.: Open-Source as Enabler of Entrepreneurship Ambitions Among Engineering Students – A Study Involving 20 Finnish Startups. In: Proceedings of the International Conference on Engineering Education 2012, Research reports 38, pp. 623–629. Turku University of Applied Sciences (2012)

Towards Understanding of Structural Attributes of Web APIs Using Metrics Based on API Call Responses

Andrea Janes, Tadas Remencius, Alberto Sillitti, and Giancarlo Succi

Free University of Bozen-Bolzano, Bolzano, Italy
{andrea.janes,tadas.remencius,alberto.sillitti,giancarlo.succi}@unibz.it

Abstract. The latest trend across different industries is to move towards (open) web APIs. Creating a successful API, however, is not easy. A lot depends on consumers and their interest and willingness to work with the exposed interface. Structural quality, learning difficulty, design consistency, and backwards compatibility are some of the important factors in this process. The question, however, is how one can measure and track such attributes. This paper presents the beginnings of a measurement framework for web APIs that is based on the information readily available both to API providers and API consumers - API call responses. In particular, we analyze the tree-based hierarchical structure of JSON and XML data returned from API calls. We propose a set of easy-to-compute metrics as a starting point and describe sample usage scenarios. These metrics are illustrated by examples from some of the popular open web APIs.

1 Introduction

More and more businesses are taking advantage of the so-called API Economy every day. Powered by the (open) web APIs, this new way of doing business [1] offers a number of exciting opportunities [2], such as getting additional value of your company's business assets and fostering innovation from third-parties [3] - all at a very low cost[1].

The number of available web APIs has been increasing in a rapid manner. For example, the registry of open web APIs at *programmableweb.com* [4] shows close to exponential growth rate (Figure 1).

In principle, a web API can be implemented both as a proprietary and as an open-source software application, yet, conceptually, it represents a middle-ground between these two models. The code of the API or at least the source of the data/services it exposes is typically hidden from the API consumers, giving protection to the API provider and maintaining its ownership of the valuable business assets. The exposed API interface, on the other hand, becomes the open

---
[1] Compared to normal effort and investment needed to enact new business initiatives, such as development of new products/features or starting marketing campaigns for reaching new markets.

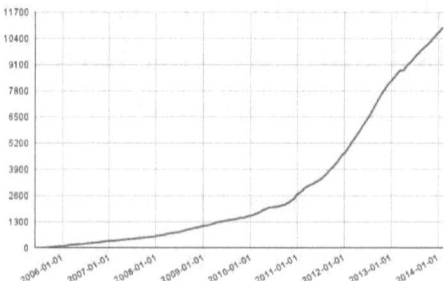

Fig. 1. Open web API growth rate, as extracted from programmableweb.com [4]

pseudo-code that is generally available [1] to third-parties to use and extend as they please. For example, consumers can combine APIs from different providers to offer completely new products with very little effort required. It's a model that follows a *win-win* approach, enabling both providers and consumers to get benefits from the exposed API while maintaining a certain degree of separation and independence from each other. Success of one side means greater chances of success for the other. This makes the risk of one of the parties abusing the work of the other much lower.

As one could expect, however, the API Economy comes with its own set of challenges and difficulties [2], [5]. One of those is how to create a successful API. This includes being able to attract API consumers and providing the right level of help and support so that they remain satisfied. Bad interface design can make the API very difficult to learn and to use, thereby hindering its adoption. Failing to maintain backwards compatibility or introducing breaking changes can not only make existing consumers abandon the API but can also prevent arrival of new ones, because of the damage to the reputation of the provider.

From the technical (development) side, implementation of web APIs is not much different from implementation of other type of software solutions. In fact, because the code of the API is a *black-box* to consumers, the way how it is actually implemented is not that important - what matters is that it works according to current expectations.

The API interface, on the other hand, becomes the visible *meta code* that really matters. Its structure has direct impact on API complexity, maintainability (i.e., providing backwards-compatibility) and easiness to learn and use. Unfortunately, there are yet no established metrics that would allow to measure and understand the APIs based on their exposed interface. Our work is aimed to facilitate progress in this direction.

In the following sections, we describe a set of relatively simple metrics that could be used as a starting point in API analysis. While it is still too early to say how beneficial these metrics could be in practice, one can already get a sense of their potential applicability to different situations. The examples provided in the paper come from our analysis of some of the popular open web APIs[2].

1.1 Web APIs

The term *web API* can be somewhat confusing to the old-school programmers who are used to the original notion of the API (i.e., as a library of functions, methods, and/or classes written for a specific programming language). In the context of the API Economy this term becomes more of a *meta* concept. The part that stays the same is that the API is still a collection of functionalities. However, it is no longer tied to a specific programming language and its purpose is to expose business assets as opposed to facilitation of reuse of the existing code. Here are the definitions that we use:

Definition 1 (API Economy). *An economy in which companies expose their (internal) business assets or services in the form of (web) APIs to third parties with the goal of unlocking additional business value.*

Definition 2 (Web API). *A software interface that exposes certain business assets to other parties (systems and/or people) over the web.*

Definition 3 (API Call). *An HTTP request to the web API and a corresponding response.*

Web APIs can generally be executed using a web browser and their output typically is in a human-readable form. In particular, REST [6] and REST-like[3] APIs that use XML or JSON format are dominating the market (Table 1).

Table 1. Most used protocols and response formats[4]

Format	APIs	%	Protocol	APIs	%
XML or JSON	8882	81.08%	REST-like	7771	70.94%
XML	6047	55.20%	SOAP	2135	19.49%
JSON	5176	47.25%	not specified	1086	9.91%
XML and JSON	2341	21.37%	JavaScript	594	5.42%
not specified	1843	16.82%			
Total APIs: 10955					

---

[2] Based on the list of popular APIs from *programmableweb.com* [4],
 http://www.programmableweb.com/apis/directory/1?sort=mashups
[3] Following loose adherence to REST.
[4] Based on data extracted from *programmableweb.com* [4]

1.2 JSON and XML Based Responses

JSON is a data format that has been gaining a lot of popularity in recent years. One of the main reasons for its quick spread is that it is very easy to transform data from an XML structure to a JSON one and visa-versa. As a result, a number of web APIs support both output formats (Table 1).

JSON and XML enable representation of hierarchical semi-structured data. As such, every response to a request could be represented as a tree. An API as whole could then be seen as a tree composed of all call trees. This is the basis of our approach to the analysis of the API structure.

1.3 Our Approach

Our current strategy is to focus solely on the response part of API calls. The structure of call requests is generally much simpler and shorter and, therefore, has less impact on the resulting API. That said, however, we do have plans for incorporating request analysis in the future.

We also consider only the structure of the response and not the actual values contained within. Analyzing actual data could have its uses, in particular when it comes to testing, but again, we felt that it was important to focus on the structural aspects of the APIs first. This has a side benefit, as it means that data does not need to be present during the analysis, eliminating information security and privacy concerns.

The results of our ongoing research can be followed and freely accessed at our website [7]. We provide an open web API implementing analysis and metric computation [8,9], as well as a web GUI interface that is built on top of that API. We encourage any interested readers to visit our site and contribute with feedback, ideas, and suggestions.

2 Measurement Framework

Our analysis of API structure is based on three directions: (1) computation of tree-based metrics, (2) computation of node-name-based (semantic) metrics, and (3) visualization of APIs and API calls.

2.1 Objectives

It is difficult to predict all possible uses for the metrics in advance. Nevertheless, we have several general objectives that serve as our guideline:

- Evaluate how difficult it is to learn and/or use the given API.
- Understand what the problems in the current design of the API are.
- Evaluate API consistency and backwards-compatibility.
- Understand how one API differs from another and what the impact of those differences is.

2.2 Metrics

As described in the previous section, at the moment we use only the structure of call responses as our analysis target. As such, when we analyze an API we include only those calls that have output in the form of XML or JSON. We also consider only the typical responses to calls, disregarding special cases, like errors and exceptions. The latter typically have a very different structure from normal responses and their inclusion into analysis is left for future research.

We do consider different call responses based on variations in request parameters, but only when there is a structural difference present (e.g., when different request parameters result in different type of data returned).

It is common for responses to contain sets of elements of the same type. The same request might result in a different number of items returned depending on the context (e.g., the user who makes the request, current amount of data, etc.). From the structural point of view we do not care how many items are returned as long as they all are of the same type. For this reason, every request is first *sanitized* by merging sibling nodes that have exactly the same structure (see Figures 2 and 3). This means that no matter the amount of actual data returned, the resulting response representation stays the same.

 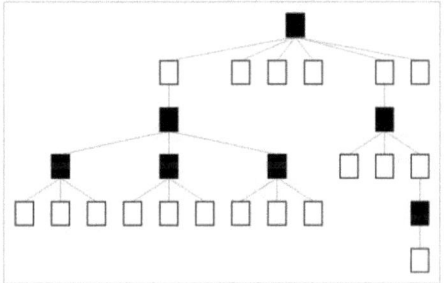

Fig. 2. JSON response and its visualization as a tree (black rectangles represent *meta* nodes). API: PayPal, API call: GET https://api.sandbox.paypal.com/v1/payments/capture/8F148933LY9388354

2.3 Basic Tree Metrics

Every tree can be characterized by some simple metrics that reflect on its dimensions and/or shape.

One simple way to look at the size of a tree is to count the *number-of-nodes* (structural elements) it contains.

A related metric is the *number-of-unique-nodes*. *Unique* in this context means having a different node name from those that were counted before. This measure shows the *richness* of structure in terms of node variety. Used in conjunction with normal node count (e.g., as a ratio) it can be employed to identify the

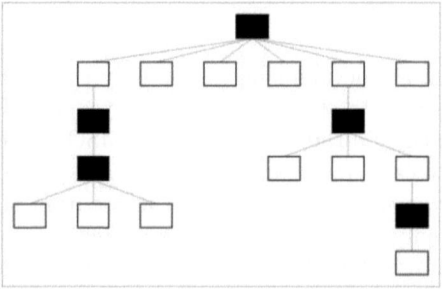

Fig. 3. Visualization of the call tree after merging matching subtrees (left), displayed with some of the call metrics (right). API: PayPal, API call: GET
https://api.sandbox.paypal.com/v1/payments/capture/8F148933LY9388354

presence of dominant node names, which might be an indication of duplicated values or overloaded naming (e.g., semantically different data parts given the same names).

Another way of looking at the size of a tree is to consider its node topology. We use three metrics for that purpose: *tree-depth*, *tree-breadth*, and *number-of-children-per-node*.

Tree-depth is the length of the path from the root of the tree to its furthest leaf.

Tree-breadth is the maximum number of nodes on one level of the tree. The level is the distance of a node from the root (i.e., the root node is at level 0, its child nodes at level 1, and so on).

When depicted visually, depth represents the height of the tree shape and breadth - the width.

The last metric we use is the *number-of-children-per-node*. It is computed as an average, though minimum and maximum values per tree might also offer valuable insights. This metric shows how much nodes tend to branch.

2.4 API Metrics

We employ two ways of computing API-wide metrics. The first one is based on the aggregation of metrics computed for each API call. This usually means computing average, minimum, and maximum values.

The second approach is to combine all call trees into one big API tree (see Figure 4 for an example of an API tree visualization) and compute the metrics defined in the previous subsection. This is performed in two steps:

1. We introduce a virtual root node and add all call trees as child nodes.
2. We merge sibling nodes that have the same structure. This way two or more calls might get combined into one subtree.

The merging step has two parameters that can be adjusted based on the desired effect. The first one determines whether to do the merge when a single

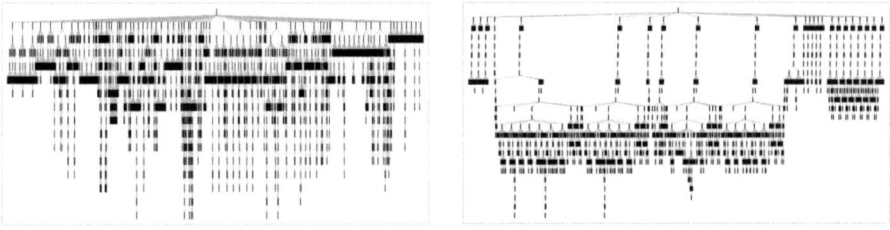

Fig. 4. Visualization of the API trees: left - *Twitter* API, right - *Bing Maps* API

element is compared to a list of elements of the same type as that element. Our default strategy is to execute the merge. This is because a single node can be seen as a list containing just one element (that node). In the situations when the multiplicity of nodes (i.e., one vs. many relationship) is considered to be important, such nodes can be left separate.

The second parameter is the merge factor. By default, we do the merge only when the structure of siblings matches exactly. However, this criteria can be relaxed using the *similarity* measure (described later in the paper). The motivation for merging similar but not exactly the same subtrees comes from the fact that some APIs have optional fields that are only returned if the appropriate values are present or if a specific parameter was added to the request.

2.5 Node-Name-Based (Semantic) Analysis

A different approach to analyzing API structure is to focus on the actual names of the nodes. These names are normally not given at random and represent semantics of the structure.

The simplest set of metrics of this type is the *count-of-each-named-node*. It can be computed for individual calls and for the whole API. The most frequent and the least frequent names are usually of most interest. Outlier values, in terms of name frequency, might indicate bad design or names that require special attention in the help manual or in the API reference.

A more advanced form of analysis is to consider which node names go together and how often. For example, one can count the occurrences of different *node pairings* when they appear as siblings (nodes that have the same parent node).

2.6 Similarity-Based Metrics

A known problem in the analysis of tree-based data is how to compare two trees. A common way of computing the similarity between trees is to use the so-called *edit-based distance* [10]. It can be expressed as the minimum number of edit operations required in order to convert one tree to the other.

Such *similarity* measure can also be used for comparing API responses, but it has one major disadvantage, besides being complex and computationally intensive. It considers the number of required edits but does not consider their location (does not account for the significance of the hierarchy). In the case of the API response structure, we feel that the differences on the higher level of the tree (closer to the root) should have more impact. At the same time, we do not need a measure with the perfect precision. It is enough that we can tell exactly when two trees are the same, and to express their similarity in a somewhat intuitive manner in other cases. For this purpose we have devised the following *similarity* metric.

We consider that the *similarity* of two trees can be described by *similarities* among their subtrees, where each subtree has the same impact, independently from its size. So, if there are 5 subtrees across both trees, the impact of each would be 20%. However, we also want to account for the names of the root nodes, otherwise two trees that have the same root names but totally different subtrees would have a *similarity* of 0. There are different possibilities here in terms of how much weight we could give to the matching of such roots. It all depends on the point of view. It could be a fixed ratio, like 33% or 50%, but that would set the fixed minimum value for trees with matching roots. Therefore we feel that a weight related to the impact of one subtree is better fitting. At this point we simply use the weight of 1.0 - the same impact as that of a single subtree. When the root names do not match the trees are considered not similar and get the score of 0. The exception to this rule are the *unnamed* nodes from JSON - elements and lists ({} and []) - they are counted as matching but are not considered to impact similarity by themselves (impact weight of 0).

The detailed steps to compute the *similarity* are shown below:

1. Check if the names of the roots match or if both are *meta* nodes. If not, the similarity is 0. Otherwise proceed to the analysis of subtrees (step 2).
2. Take the tree that has fewer subtrees (T_S; L_{T_S} - the number of subtrees of T_S). For each subtree i (i = 1, ..., L_{T_S}) of T_S compute the similarity with subtrees of the other tree (T_L; L_{T_L} - the number of subtrees of T_L). Memorize the highest similarity S_i (take the first one if multiple *similarities* match; stop the checks if a *similarity* of 1.0 - an exact match - is found) and remove the corresponding subtree of the tree T_L from further computations, unless no similar subtrees are found (when maximum similarity S_i was 0). In the latter case just memorize the value (0) and proceed to the next subtree. Once all subtrees from T_S are processed, combine the memorized impacts (step 3).
3. If the names of the roots matched, compute the final *similarity* score thusly:

$$S = \left(1 + 2 \cdot \sum_{i=0}^{L_{T_S}} S_i\right) \div (1 + L_{T_S} + L_{T_L}) \ . \tag{1}$$

Otherwise (in the case of *meta* nodes), the roots are considered to have no impact:

$$S = \left(2 \cdot \sum_{i=0}^{L_{T_S}} S_i\right) \div (L_{T_S} + L_{T_L}) \ . \tag{2}$$

The resulting value of the *similarity* ranges from 0.0 (not similar) to 1.0 (exactly the same).

Besides using this metric directly to compare two API calls and as a criteria for merging sibling nodes (as described in the previous sections), we also use it to build the API *similarity matrix* (see Figure 5 for an example).

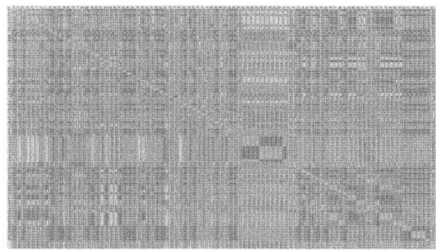
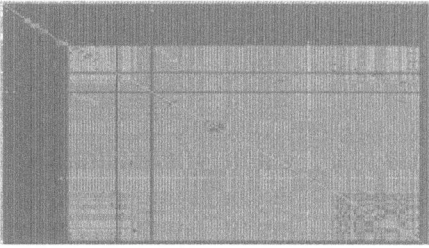

Fig. 5. Color-coded *similarity matrices* for the *Twillio* (left) and *Wikipedia* (right) APIs. Colors have the following mapping to the *similarity scores*: red - 0 (not at all similar), yellow - (0.0-0.5] (somewhat similar), green - (0.5, 1.0] (very similar)

API *similarity matrix* is the matrix containing *similarities* between every two calls of the API. Such matrix can be used to determine which calls could be grouped together (e.g., in the API manual, for better comprehension) or to verify if existing groupings are consistent.

Table 2. Examples of API *similarity scores* computed on different APIs

API	Number of Calls	Similarity Score
PayPal	19	0.2966
Twillio	94	0.2263
Twitter	78	0.1104
NASA	9	0.3723
Wikipedia	154	0.1726
BingMaps	24	0.9041
Salesforce	26	0.0644

We call the average of all similarities from the matrix, computed excluding self-similarities (i.e., a call compared to itself), the *similarity score* of the API. It represents the general consistency of the response structure within that API (see Table 2 for examples).

The *similarity* metric can also be used to analyze version evolution and the amount of structural changes (impact on backwards-compatibility) for an API call. It can also be applied for comparison of two different APIs (Figure 3).

Table 3. Examples of *similarity scores* computed between pairs of APIs

	PayPal	Twitter	NASA	Twillio	Wikipedia	Salesforce
PayPal	0.2966	0.0145	0	0.0003	0	0.0223
Twitter	0.0145	0.1104	0.0035	0.0033	0.004	0.0102

3 Conclusion

Although the metrics presented here have been tested on a set of open web APIs and appear to highlight certain differences in the call response structure, their practical value has yet to be properly evaluated.

Nevertheless, we hope that this paper will give ideas and motivation to other researchers and practitioners, and help in establishing a deeper body of knowledge regarding web APIs, their design, and their evolution throughout the API lifecycle.

References

1. Gat, I., Remencius, T., Sillitti, A., Succi, G., Vlasenko, J.: API Economy: Playing the Devil's Advocate. Cutter IT Journal 26(9), 6–11 (2013)
2. Gat, I., Succi, G.: A Survey of the API Economy. Cutter Consortium Agile Product & Project Management Executive Update 14(6) (2013), http://www.cutter.com/content-and-analysis/resource-centers/agile-project-management/sample-our-research/apmu1306.html
3. Clark, J., Clarke, C., De Panfilis, S., Granatella, G., Predonzani, P., Sillitti, A., Succi, G., Vernazza, T.: Selecting components in large cots repositories. Journal of Systems and Software 73(2), 323–331 (2004)
4. programmableweb.com, http://www.programmableweb.com/
5. Remencius, T., Succi, G.: Tailoring ITIL for the Management of APIs. Cutter IT Journal 26(9), 22–29 (2013)
6. Fielding, R.T.: Architectural styles and the design of network-based software architectures. Ph.D. dissertation, University of California, Irvine (2000)
7. apiwisdom.com, http://www.apiwisdom.com/
8. Scotto, M., Sillitti, A., Succi, G., Vernazza, T.: A relational approach to software metrics. In: Proceedings of the 2004 ACM Symposium on Applied Computing, pp. 1536–1540. ACM (2004)
9. Scotto, M., Sillitti, A., Succi, G., Vernazza, T.: A non-invasive approach to product metrics collection. Journal of Systems Architecture 52(11), 668–675 (2006)
10. Bille, P.: A survey on tree edit distance and related problems. Theoretical Computer Science 337(1), 217–239 (2005)

Open Source Mobile Virtual Machines: An Energy Assessment of Dalvik vs. ART

Anton B. Georgiev, Alberto Sillitti, and Giancarlo Succi

Free University of Bozen Bolzano (UNIBZ) Piazza Domenicani 3
39100 Bolzano, Italy
{anton.georgiev}@stud-inf.unibz.it,
{Alberto.Sillitti,Giancarlo.Succi}@unibz.it

Abstract. Dalvik Virtual Machine is Open Source Software and an important part of the Android OS and its better understanding and energy optimization can significantly contribute to the overall greenness of the mobile environment. With the introduction of the OSS solution, named Android Runtime (ART) an attempt of performance and energy consumption optimization was made. In this paper we investigate and compare the performance of the Dalvik virtual and ART from energy perspective. In order to answer our research questions we executed a set of benchmarks in identical experimental setup for both runtimes, while measuring the energy spent and percentage of battery discharge. The results showed that in most of the use case scenarios Ahead-Of-Time compilation process of ART is overall more energy efficient than the Just-In-Time one of Dalvik.

1 Introduction

Dalvik was the virtual machine that Google created to use in Android, its mobile Operating System. For 7 years it evolved alongside with other modules of Android. Dalvik made use of Just-In-Time compiler as a suitable solution for devices with space limitations, characteristic of the first generation devices equipped with Android OS. With the growth of computational power, storage space, display size and embedded sensors, these limitations become less restrictive. This enabled the Operating System developers to seek performance optimizations and better UI responsiveness by introducing new virtual machine. ART stands for Android runtime and it was introduced as a new feature available in version 4.4 of the Android Operating System. It was an optional feature in that release because it was in experimental phase, and its inclusion was triggered by the need of having feedback from both users and developers.

In this paper we designed and executed an experiment to analyze how the new ART affects the overall energy consumption. We based our study on the execution of several benchmarks.

The rest of this paper is organized as follows: section 2 introduces Android OS virtual machine; section 3 comments related work; section 4 describes the methodology and environment created to carry out our experimentation; section 5 presents and analyzes the data collected during the benchmarks' executions and section 6 identifies directions for future work and draws the conclusions.

2 Android OS and Its Virtual Machines

The Android Operating System is an open source operative platform for managing mobile devices. Its open nature allows researchers to analyze its structure and internal organization, facilitating a richer analysis and offering better development possibilities [1,2].

Android is organized in different layers to place the services and applications according to specific requirements and necessities. It is based upon a customized version of the Linux kernel, which acts as the abstraction layer between the hardware and the software. On top of the kernel there is a set of Open Source C/C++ libraries. In the topmost part of the stack the Android apps reside [3]. These apps are programmed in Java programming language. Such programs are compiled into .dex (Dalvik Executable) files, which are in turn zipped into a single .apk file on the device. Android's dex files is created by automatically translating compiled applications written in the Java programming language; the Java compiler (javac) then produces Java bytecode. The dex compiler (dex) then translates java bytecode to a Dalvik bytecode; and Dalvik bytecode is then executed by a virtual machine called Dalvik (DVM) [3].

With the advancement of the mobile devices, it was reached a level where the storage space is relatively big Android developers decided to introduce new virtual machine called Android runtime (ART). It is an alternative of the Just-In-Time compiler, which was part of the Android OS since version 2.2 [4].

With the release of ART Google aimed to improve the GUI responsiveness, to shorten the loading time of the mobile applications and to optimize the execution time and energy consumption of its system. However it has not been discussed what would be eventual implication at user level, for instance an additional use of resources or and additional energy tall

The goal of this work is to analyze the performance of Dalvik and ART from energy point of view.

With the purpose to reach this goal, we defined two research questions:

R.Q.1 Are there any differences between Dalvik and ART in terms of energy efficiency?

R.Q.2 How big is the difference between the two FOSS technologies from the energy perspective?

In order to answer our research questions we executed a set of benchmarks in the same experimental setups for both Dalvik and ART, measuring the energy spent and percentage of battery discharge.

3 Related Work

The Open Source nature of the Android OS in combination with mobile area triggered the interest of many enthusiasts to benchmark and measure the performance of both Dalvik and ART. Since ART was introduced in version 4.4 of the Android OS there are just a few articles [5, 6]. Based on their results one can conclude that even in experimental phase, there are use cases (e.g. floating point calculations) where ART performs better than Dalvik.

In contrast with ART, Dalvik was part of Android since the beginning. During the years of Android OS evolution there were series of studies analyzing its virtual machine. Some of them offer a special focus on assessing the performance of Java and C implementations, conduct a comparison between them [7, 8]. Other work [9] presented a review of the energy consumption of an Android implementation in comparison with the same routine in Angstrom Linux.

In a wider scope there are several research publications than were found during our literature survey dealing with the problem of power consumption in a smartphone. The work of Olsen and Narayanaswami [10] presents and Operating System's power management scheme, named PowerNap, which aims to improve the battery life of mobile device by putting the system in more efficient low power state. The authors of another research publication [11] analyzed the power consumption of a mobile device and measured not only the overall energy spent, but also the exact break down by the device's main hardware components. In another track, analyzing component specific energy characterization is an aim for the authors [12], which used a software tool named CharM to survey and collect parameters, exposed by the Operating System. The presented results showed that CPU, the OLED display and the WiFi interface are the elements that are taking the highest tall from the device battery when stressed.

For the purposes of our open source experimental setup, we needed a state of the art analysis of the mobile software based energy measurement. After surveying the literature in the area we discovered a set of applications which implements profiling techniques that analyze the source code. With these tools at hand you can obtain very detailed information about what are the routines and system calls that have a major contribution to the energy utilization [13]. Building an energy model using hardware measurement is another approach that is used in this area [14, 15, 16, 17, 18].

4 Data Collection

4.1 Benchmarks

To explore all the differences between the two runtimes we selected several Benchmarking apps from Google play. These software products will guaranteed a fair comparison, because they are executing the same set of instruction in identical sequence each time they perform their benchmark. The benchmark applications utilized on our experimentation were:

- *AndEBench* [19] – it is an app providing standardized method, which is industry-accepted for evaluating Android OS performance.
- *Antutu Benchmark* [20] – is a benchmark tool, which includes – CPU, GPU, RAM, I/O tests with the option to compare your results with popular Android OS devices
- *GFXBench 3.0* [21] – is a comprehensive benchmark and measurement tool, with cross-platform and cross-API capabilities. It measures graphics performance alongside with render quality and long-term performance stability of the device. We selected to include two of the available Low-Level tests – basic ALU and its 1080p off-screen variation.
- *Pi Benchmark* [22] – calculates Pi up to the 2.000.000 digits of precision. For our experiment we were execution this benchmark with load of 500.000 digits.
- *Quadrant Standard Edition* [23] – is an application, which requires Internet connection to compute the results and doesn't work on device without GPU. It stresses the CPU, GPU and performs I/O operations.
- *RL Benchmark: SQLite* [24] – determines how much time you need to process a number of SQL queries.
- *Vellamo Mobile Benchmark* [25] – evaluates the web browser performance of the device including scrolling and zooming, 3D graphics, video and memory read/write performance.

With these benchmarks we also made sure that we exercise different components that may impact the execution and the overall user experience while using an app.

4.2 Experimental Setup

Our experimental setup used two software tools to collected information regarding the energy consumption of the executed benchmarks:

- *A Custom version of PowerTutor* [13] to measure the energy spent in Joules. The samples were collected after every single benchmark execution.
- *CharM* [charm] – to read the battery percentage spent. CharM is a tool that surveys battery discharge cycle. It collects the date in a background process, and is implemented with negligible energy footprint. We used it to collect date more relevant to the end user, as one would normally check the remaining battery percentage on his device. The readings were acquired after the complete cycle of benchmark executions was over.

Before each measurement phase we were resetting the PowerTutor and CharM instances to assure that the benchmark application will start its energy consumption from 0. The battery of the phone was fully charged and left to cool down to assure that the settings for each test are as identical as possible. For reproducibility purposes we repeated each test 30 times.

Our test bed options were limited, because ART was experimentally introduced only for Android 4.4 compatible devices. We selected a Google® HTC Nexus 4™ device and all the results presented in this paper are according to its capabilities and specification. The phone has the latest stock Android version (4.4.2), without any account information in order to prevent background synchronization with remote servers, which can affect our measurements. The device is equipped with a quad-core Snapdragon S4 Pro APQ8064 processor running at 1.5 GHz, 2 GB RAM and 2100 mAh battery.

5 Data Analysis

5.1 Data Interpretation

After executing all the benchmark series we collected more than 500 samples for both virtual machines. Table 1 shows the average energy consumption in Joules of 30 cycles for each benchmark. Column two contains readings associated with Dalvik virtual machine, while column three contains the ART ones.

Table 1. Summary of the energy spent per benchmark application in Joules

Benchmark Name	Dalvik	ART
AndEBench [b1]	88.36 J	115.18 J
Antutu Benchmark [b2]	265.66 J	265.18 J
GFXBench 3.0 [b3]	90.50 J	90.50 J
Pi Benchmark [b4]	217.40 J	92.68 J
Quadrant Standard Edition [b5]	77.56 J	67.62 J
RL Benchmark:SQLite [b6]	51.46 J	29.30 J
Vellamo Mobile Benchmark [b7]	293.18 J	284.10 J

The energy spent to complete each computational job varies largely due to the nature of the selected benchmarks. Some of them like Antutu Benchmark [20] and Vellamo Mobile Benchmark [25] considerably higher energy consumption than the rest of the experiments.

By the graphics shown in Figure 1, with the exception of AndEBench, ART performed equally or better in comparison with Dalvik. We have the biggest difference in the Pi Benchmark where the newer virtual machine was more than two times efficient in comparison with its predecessor.

In Table 2 is summarized the average of the percentage that was discharge from the battery during one benchmark series (i.e. the execution of a benchmark for 30 times). The readings were collected using CharM application.

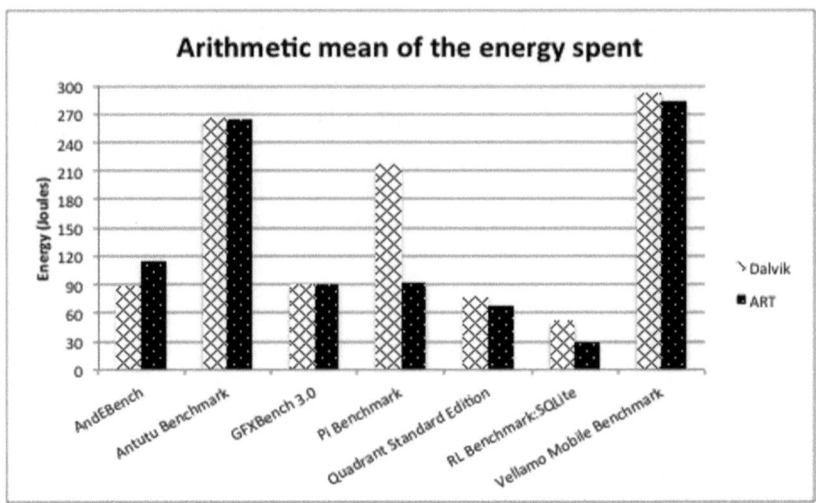

Fig. 1. Arithmetic mean of the energy spent

Table 2. Summary of the percentage of battery discharge per benchmark application

Benchmark Name	Dalvik	ART
AndEBench	32 %	45 %
Antutu Benchmark	57 %	57 %
GFXBench 3.0	30 %	30 %
Pi Benchmark	55 %	36 %
Quadrant Standard Edition	35 %	31 %
RL Benchmark:SQLite	11 %	6 %
Vellamo Mobile Benchmark	74 %	61 %

The plots of the remaining battery level after each benchmark execution per selected experiment can be seen on Figure 2. The figure presents information about the battery level indicator that the end user will see on his screen after the completion of each experimental task.

5.2 Research Questions Revisited

R.Q.1 Are there any differences between FOSS virtual machines Dalvik and ART in terms of energy efficiency?

After analyzing the collected results it is evident that in almost all cases we have difference in the energy spent between the two virtual machines. The only benchmark that we had exactly the same reading was *GFXBench 3.0* with 90.50 Joules for both Dalvik and ART.

R.Q.2 How big is the difference between the two FOSS technologies from energy perspective?

Readings in Table 1 show that out of seven benchmarks only in one, namely *AndEBench*, we have advantage of 26.82 Joules for the Dalvik virtual machine. In the remaining 6 benchmarks, excluding *GFXBench 3.0*, we always have better energy efficiency for the ART virtual machine with average savings from 0.48 Joules (for *Antutu Benchmark*) to 124.72 Joules (for *Pi Benchmark*).

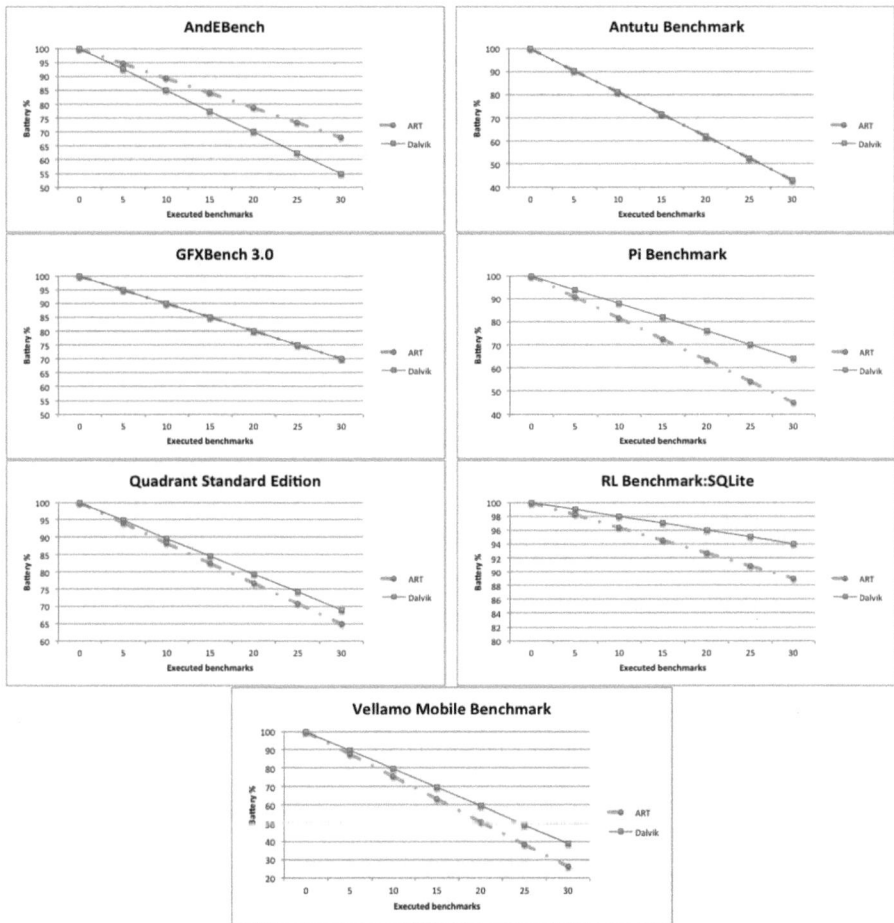

Fig. 2. Battery discharge level after benchmark execution in Dalvik and ART

6 Future Work and Conclusions

6.1 Directions for Future Work

Possible continuation of the work presented in this paper is to replicate the benchmark executions on different test beds. Collecting samples not only from smart phones, but also from tablets, smart TVs and others Android embedded device will deepen our understanding of the energy performance in the Operating System. Reproducing the described experiments using hardware measurement tools instead of software will provide additional verification of the previously generated data. Moreover we also foresee to increase the granularity of the collected measurements [26], so in the future we can analyze in greater details the relationship between the code execution and the corresponding energy consumption.

6.2 Conclusions

In this work we presented and analyzed data collected from set of selected benchmarks in order to analyze how the new Android runtime affects the overall energy consumption of the device under test. These benchmarks were selected with the goal of stressing different components that may impact the execution and the overall user experience while using an app. The results showed that replacing a Just-In-Time compiler with virtual machine with Ahead-Of-Time compilation process would optimize the energy utilization for five out of seven scenarios exercised using our experimental setup.

Optimizing the performance of a mobile device and by doing so reducing the energy consumption is of a great importance for the future development of the mobile environment. Having a strong requirement for autonomy provokes not only the mobile hardware specialists, but also software developers to seek for a solution that will increase the battery life. If they omit the energy consumption factor while designing their products, they could not only decrease the user satisfaction, but in a long term they might significantly contribute to environmental pollution by increasing the battery garbage.

References

[1] Di Bella, E., Sillitti, A., Succi, G.: A multivariate classification of open source developers. Information Sciences 221, 72–83 (2013)
[2] Kovács, G.L., Drozdik, S., Succi, G., Zuliani, P.: Open source software for the public administration. In: Proceedings of the 6th International Workshop on Computer Science and Information Technologies (2004)
[3] Ehringer, D.: The Dalvik Virtual Machine Architecture (March 2010), http://davidehringer.com/software/android/The_Dalvik_Virtual_Machine.pdf (retrieved November 4, 2013)
[4] Dalvik (software). In Wikipedia (software) (December 1, 2013), http://en.wikipedia.org/wiki/Dalvik_(software) (retrieved December 2, 2013)
[5] Toombs, C.: Meet ART, Part 2: Benchmarks - Performance Won't Blow You Away Today, But It Will Get Better (November 12, 2013), http://www.androidpolice.com/2013/11/12/meet-art-part-2-benchmarks-performance-wont-blow-away-today-will-get-better/ (retrieved December 1, 2013)

[6] Toombs, C.: Meet ART, Part 3: Battery Life Benchmarks - Not Good, But Not Too Bad (January 22, 2014), http://www.androidpolice.com/2014/01/22/meet-art-part-3-battery-life-benchmarks-not-good-but-not-too-bad/ (retrieved January 26, 2014)

[7] Cargill, D.A., Radaideh, M.: A practitioner report on the evaluation of the performance of the C, C++ and Java compilers on the OS/390 platform. In: Proceedings of the IEEE International Symposium on Performance Analysis of Systems and Software, pp. 40–45 (2000)

[8] Corsaro, A., Schmidt, D.: Evaluating Real-Time Java Features and Performance for Real-time Embedded Systems. In: Proceedings of the Eighth IEEE Real-Time and Embedded Technology and Applications Symposium, pp. 90–100 (2002)

[9] Kundu, T.K., Kolin, P.: Android on Mobile Devices: An Energy Perspective. In: Proceedings of the 2010 10th IEEE International Conference on Computer and Information Technology, pp. 2421–2426 (2010)

[10] Olsen, C.M., Narayanaswami, C.: PowerNap: An efficient power management scheme for mobile devices. IEEE Trans. on Mobile Computing 5(7), 816–828 (2006)

[11] Carroll, A., Heiser, G.: An analysis of power consumption in a smartphone. In: USENIX Annual Tech. Conf., pp. 21–34 (2010)

[12] Corral, L., Georgiev, A.B., Sillitti, A., Succi, G.: A Method for Characterizing Energy Consumption in Android Smartphones. In: Proceedings of the 2nd International Workshop on Green and Sustainable Software (GREENS 2013), in connection with ICSE 2013, pp. 38–45 (2013)

[13] Zhang, L., Tiwana, B., Qian, Z., Wang, Z.: Accurate online power estimation and automatic battery behavior based power model generation for smartphones. In: 8th Intl. Conf. on HW/SW Codesign and System Synthesis, pp. 105–114 (2010)

[14] Wonwoo, J., Chulko, K., Chanmin, Y., Donwon, K., Hojung, C.: DevScope: a nonintrusive and online power analysis tool for smartphone hardware components. In: 8th Intl. Conf. on HW/SW Codesign and System Synthesis, pp. 353–362 (2012)

[15] Yoon, C., Kim, D., Jung, W., Kang, C., Cha, H.: AppScope: application energy metering framework for Android smartphone using kernel activity monitoring. In: USENIX Annual Technical Conference (2012)

[16] Pathak, A., Hu, Y., Zhang, M.: Where is the energy spent inside my app? Fine grained energy accounting on smartphones with Eprof. In: 7th ACM European Conference on Computer Systems, EuroSys 2012, pp. 29–42 (2012)

[17] Corral, L., Sillitti, A., Succi, G., Strumpflohner, J., Vlasenko, J.: DroidSense: A mobile tool to analyze software development processes by measuring team proximity. In: Furia, C.A., Nanz, S. (eds.) TOOLS 2012. LNCS, vol. 7304, pp. 17–33. Springer, Heidelberg (2012)

[18] Corral, L., Sillitti, A., Succi, G., Garibbo, A., Ramella, P.: Evolution of mobile software development from platform-Specific to web-Based multiplatform paradigm. In: Proceedings of the 10th SIGPLAN Symposium on New Ideas, New Paradigms, and Reflections on Programming and Software, pp. 181–183. ACM (October 2011)

[19] EEMBC, AndEBench (Version 1605) (2014), https://play.google.com/store/apps/details?id=com.antutu.ABenchMark&hl=en (retrieved)

[20] AnTuTu, Antutu Benchmark (Version 4.1.7), https://play.google.com/store/apps/details?id=com.eembc.coremark (retrieved)

[21] Kishonti Ltd. GFXBench 3.0 3D Benchmark (Version 3.0.4) (2014), from `https://play.google.com/store/apps/details?id=com.glbenchmark.glbenchmark27` (retrieved)
[22] Markovic, S.D.: PI Benchmark (Version 1.0) (2014), `https://play.google.com/store/apps/details?id=rs.in.luka.android.pi` (retrieved)
[23] Aurora Softworks, Quadrant Standard Edition (Version 2.1.1) (2014), `https://play.google.com/store/apps/details?id=com.aurorasoftworks.quadrant.ui.standard` (retrieved)
[24] RedLicense Labs, RL Benchmark: SQLite (Version 1.3) (2014), `https://play.google.com/store/apps/details?id=com.redlicense.benchmark.sqlite` (retrieved)
[25] Qualcomm Connected Experiences, Inc., Vellamo Mobile Benchmark (Version 2.0.3) (2014), `https://play.google.com/store/apps/details?id=com.quicinc.vellamo` (retrieved)
[26] Scotto, M., Sillitti, A., Succi, G., Vernazza, T.: A non-invasive approach to product metrics collection. Journal of Systems Architecture 52(11), 668–675 (2006)

Improving Mozilla's In-App Payment Platform

Ewa Janczukowicz[1,2], Ahmed Bouabdallah[2], Arnaud Braud[1],
Gaël Fromentoux[1], and Jean-Marie Bonnin[2]

[1] Orange Labs, Lannion, France
{ewa.janczukowicz,arnaud.braud,gael.fromentoux}@orange.com
[2] Institut Mines-Telecom / Telecom Bretagne,
Université européenne de Bretagne, Cesson sévigné, France
{ahmed.bouabdallah,jm.bonnin}@telecom-bretagne.eu

Abstract. Nowadays, an in-app payment mechanism is offered in most existing mobile payment solutions. However, current solutions are not flexible and impose certain restrictions: users are limited to predefined payment options and merchants need to adapt their payment mechanisms to each payment provider they use. Ideally mobile payments should be as flexible as possible to be able to target various markets together with users' spending habits. Mozilla wants to promote an open approach in mobile payments by offering a flexible, easily accessible solution. This solution is analyzed, its shortcomings and possible improvements are discussed leading to an original proposal.

1 Introduction

Smartphones have changed the way mobile payments work. Marketplaces with applications have become an essential element of the mobile payment ecosystem. They have changed users' spending patterns and got especially specialized in micro-payments [1]. There are multiple application stores that offer in-app payment functionalities, like Google's or Apple's solutions. However they are mostly wall-gardened, so clients and developers need to have an account set up with the imposed payment provider. The system is easier to control since there are no unauthorized third parties, but at the same time it becomes very limited.

PaySwarm and Mozilla have chosen a more open approach, in order to implement platforms based on open standards and accessible to multiple payment providers. So far some limiting implementation choices are imposed, but these projects are still under development. Mozilla's idea of a payment platform seems to be the most open and flexible. This approach is beneficial for new and emerging markets, since different payment methods can be introduced.

This paper focuses on Mozilla's payment solution. Firstly, it is presented and analyzed. Secondly, its limits and possible improvements are discussed. Finally a solution is proposed and analyzed.

2 Mozilla's In-App Payment Platform

In-app payments are supported by Mozilla, that encourages providing a possibility of previewing an app or installing its basic version for free [2]. It also gives the possibility of implementing different marketplaces and working with different payment providers, thus it would be possible to target various markets and to address the needs of all users no matter the payment method [3].

Mozpay is a payment solution implemented in Firefox OS v.1.0 [4, 5]. Mozilla offered a WebPayment API that, via the mozpay function, allows web content to perform a payment [6]. Figure 1 shows the existing call flow.

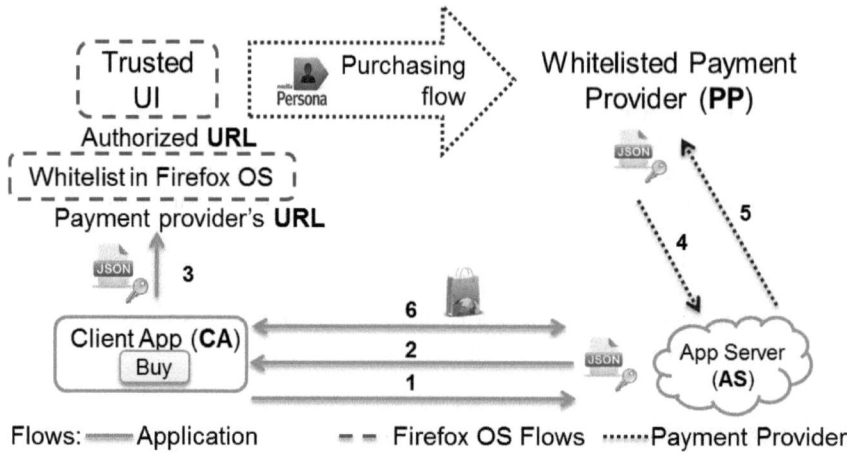

Fig. 1. Mozpay based call flow

PP implements the WebPaymentProvider API [7]. The payment flow is managed from **PP**'s server inside the trusted user interface (UI), limited to whitelisted domains. The whitelist is preregistered in the user agent and is controlled by Mozilla or whoever builds the OS. So far there is only one **PP**, created by Mozilla [2].

AS contains all application logics, manages the payment token and assures delivering goods to the user. It is assumed that to set up payments, a developer is already registered within **PP** (e.g. Firefox Marketplace Developer Hub), they have exchanged information like: financial details, application key and secret.

CA allows buying digital goods as one of its features.

The **payment token** contains all the essential information concerning a good being purchased. It is sent between all three parties throughout the payment process.

The purchase flow given below is based on Mozilla's payment provider example.

1. A user by clicking the "Buy" button requests a payment token from **AS**.
2. **AS** generates and signs the payment token, later sends it to **CA**.
3. **CA** forwards the token by calling the mozpay API [3]. If **PP** is whitelisted a trusted UI is opened, the user authenticates and the purchasing flow starts.

4. Postback (success) or chargeback (error) are tokens with additional fields (e.g. transaction ID) that inform **AS** about the payment result.
5. When **AS** receives a postback (or a chargeback) it acknowledges it.
6. In case of successful payment the purchased good is sent to the user.

The mozpay function has been proposed to be abandoned due to too rigid end-to-end transaction flows by imposing the payment token mechanism. Exposing payment provider primitives was suggested as an improvement. Payment providers would manage their own payment flows by providing JavaScript files to developers and by using a trusted UI with access restricted to whitelisted domains [8]. It can be seen on Figure 2 where the calls 1 and 2 from Figure 1 are replaced by a JS file.

3 Proposed Solution

In existing solution in order to access a trusted UI payment providers had to be whitelisted by Mozilla. It was impossible to add a new one between the Firefox OS versions. The improvement of exposing payment primitives does not solve this issue, since the access to trusted UI remains restricted to whitelisted domains [8].

A certification mechanism is a possible improvement that could replace a predefined whitelist. The solution is presented in Figure 2.

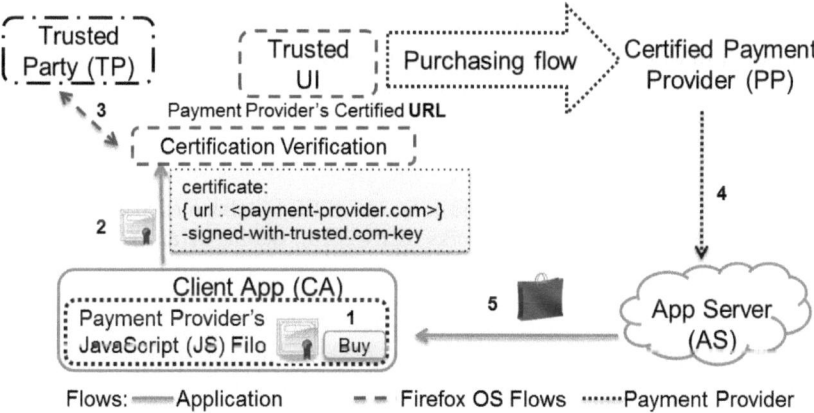

Fig. 2. Proposed solution

Instead of the mozpay method, **PP**'s JS file is included within **CA**. **PP** also provides a certificate with a URL needed to launch the trusted UI. When the Firefox OS receives a request to start the payment, it calls **TP** and verifies the certificate. If the verification is successful it uses the provided URL to open the trusted UI and the payment process begins. We assume there are no attacks on application integrity.

The proposed architecture allows changing the list of authorized payment providers without the need of redistributing the whole operating system every time a new player enters the business value chain. Instead of a central entity that controls the whitelist, there may be several trusted parties. As a result the number of payments providers would increase, so they would compete and transaction fees would become more beneficial. This also gives clients and app developers more freedom to choose a payment option. More universal system would give the possibility of efficiently targeting specific markets and clients' spending habits. There are a lot of factors to consider: different type of clients (illiterate, cash-challenged, without credit cards) and national regulations (taxes, currency). Additionally, well-known, open standards would facilitate the development process.

The drawback of the solution is that payment providers would need to adapt their flows in order to assure certificate management. Security aspects need to be studied, although an advantage of certificates is that they are widely implemented and trusted.

4 Conclusion

In-app payments are used more often but the widely used application stores or payment providers have implemented a walled-garden approach. Mozilla wants to change the way in-app payments work by offering a platform that is open and that targets new markets while not imposing strict business models. The version implemented so far has several limits. One of the biggest limitations is a whitelist of authorized payment providers that is currently shipped with the devices. The proposed solution solves this problem by offering a certification system that would manage payment providers. As a result Mozilla's solution can become more flexible and be able to meet most of participating players' requirements.

References

[1] Copeland, R.: Telco App Stores – friend or foe? In: IEEE 14th International Conference on Intelligence in Next Generation Networks (ICIN 2010), Berlin, Germany, October 11-14 (2010)
[2] https://developer.mozilla.org/docs/Mozilla/Marketplace/Marketplace_Payments (accessed November 27, 2013)
[3] https://hacks.mozilla.org/2013/04/introducing-navigator-mozpay-for-web-payments/ (accessed November 28, 2013)
[4] https://developer.mozilla.org/en-US/Firefox_OS (accessed November 29, 2013)
[5] Janczukowicz, E.: Firefox OS Overview. Telecom Bretagne Research Report RR-2013-04-RSM (November 2013)
[6] https://wiki.mozilla.org/WebAPI/WebPayment (accessed November 28, 2013)
[7] https://wiki.mozilla.org/WebAPI/WebPaymentProvider (accessed November 28, 2013)
[8] https://groups.google.com/forum/#!msg/mozilla.dev.webapi/cyk8Nz4I-f4/5er6JojC3TsJ (accessed November 13, 2013)

A Performance Analysis of Wireless Mesh Networks Implementations Based on Open Source Software

Iván Armuelles Voinov, Aidelen Chung Cedeño, Joaquín Chung, and Grace González

Research Center for Information and Communication Technologies,
University of Panama, Republic of Panama
{iarmuelles,achung,jchung,ggonzalez}@citicup.org

Abstract. Wireless mesh networks (WMNs) have emerged as a promising technology, capable of provide broadband connectivity at low cost. Implementations based on Open Source Software of these networks offer advantages for providing broadband networking communications in scenarios where cabling is too expensive or prohibitive such as rural environments. In this paper we evaluate the performance of small scale wireless mesh WMN routing protocols for WMNs: B.A.T.M.A.N. Advanced and the 802.11s standard. We also compare an OpenFlow controller implemented over the WMN, verifying their bandwidth, datagram loss and jitter.

Keywords: Open Source Software for research and innovation, Wireless Mesh Networks, OpenFlow, OpenWRT, network performance.

1 Introduction

Providing telecommunication services to difficult access areas (such as rural environments) is still difficult due to the lack of appropriate or inexpensive infrastructure. In this context, Wireless Mesh Networks (WMN) are an attractive solution for these scenarios due to their lower deployment costs and ease of expansion. WMNs could be used in community networks, home networking, video surveillance and emergency/disaster situations [1]. However, WMNs face several restrictions such as low-end equipment, single wireless channel and interferences, which degrade the overall performance of the network, and impose many drawbacks to even current and standard Internet's services. In this paper we evaluate the performance of WMN implementations based on Open Source Software for multimedia service transport. In our case, WMNs based on layer 2 routing protocol raise as a suitable and optimal connectivity choice, as any layer 3 addressing protocol could be used on top, either IPv4 or IPv6. Hence, we compare IEEE 802.11s standard [2] against Better Approach To Mobile Adhoc Networking Advanced protocol (B.A.T.M.A.N.) [3]. Also, we compare these two protocols with the OpenFlow protocol [4], which is used to control the forwarding tables of switches, routers and access points from a remote server, leveraging innovative services over the network such as access control, network virtualization, mobility, network management and visualization. The paper is organized as follows:

in Section 2, we describe the implementation of our testbed. In Section 3, we describe the test and show the results obtained. Finally, in Section 4, we describe our conclusions.

2 Testbed Implementation

The testbed consisted of a small scale WMN composed by four wireless routers. The experiments were conducted inside a small laboratory, because this indoor environment is very similar to the conditions of a real home networking scenario (Fig 1). All the wireless routers used in this testbed were TP-Link TL-WR1043ND v1.8, with four LAN ports of 1 Gbit/s Ethernet, one WAN port of 1 Gbit/s Ethernet, and one 802.11n wireless interface that works in the 2.4 GHz frequency band. We replaced the firmware of the wireless routers with the open firmware OpenWRT [5], which is a very well-known Linux distribution for embedded devices. OpenWRT supports WMN with routing protocols like OLSR, B.A.T.M.A.N. and the new standard for WMN networks 802.11s. For the experiments we used the OpenWRT Backfire 10.03.1, which comes with 802.11s support by default. To evaluate B.A.T.M.A.N. Advanced, was necessary to install the batman-adv package. For the experiments with OpenFlow we compiled OpenWRT with a package called Pantou that supports the OpenFlow protocol. The OpenFlow controller used in the experiments was POX, which is a controller based on NOX for rapid deployments of SDNs using Python. A list of all the components of the testbed and used tools based on Open Source Software is shown in the table 1.

3 Performance Evaluation

Our experiment had three scenarios, the first one was a WMN composed by four MP configured with the 802.11s standard. The second scenario was the same four MP but using the batman-adv protocol. The last one was the four MP connected to an OpenFlow controller using an out of band network for control and a data network , i.e, a wired network for the control signaling, due to a limitation of the hardware selected . The tests were conducted in a low interference environment, which maintains the optimum conditions for VoIP traffic (packet loss should not exceed 1%, the maximum delay should <150 ms and jitter must be kept below 20 ms). For our study we took measurements using Iperf either in UDP and TCP mode. Every UDP test was maintained for 300 seconds and repeated ten times, while the TCP test where maintained for 180 seconds and repeated five times. Every test was made for one hop, two hops, three hops and four hops, for both scenarios.

In the *UDP* test we found that for one and two hops the maximum *throughput* was higher in the Batman-adv scenario. This is because in the Batman-adv implementation, the wireless interfaces synchronize at 802.11n (transmission rates up to 300 Mbit/s). On the other hand, the implementation of 802.11s only synchronizes at

Fig. 1. WMN implementation

Table 1. Tools based on Open Source Software for WMN implementation and their evaluation

Name	Supported Platform	Features
OpenWRT	Wide variety of wireless routers	Allows you to customize the applications on the wireless router. It implements routing protocols such as OLSR and BATMAN. It can be adapted to work with IPv6 and supports 802.11s. OpenWrt is the framework to build an application without having to build a complete firmware around it.
Pantou	Linksys and TP-Link Wireless Router	Implementation of the OpenFlow protocol for the OpenWRT firmware
NOX/POX	Linux	OpenFlow controller based on Python and C++
Mininet	Linux	Allows you to create scalable software defined networks within a single PC.
Insider	Linux and Windows	Locate wireless networks and measures the intensity of their signals.
Iperf	Linux and Windows	Creates TCP and UDP data flows to measure the behavior of the network with respect to some QoS parameters.
Wireshark	Linux and Windows	Protocol analyzer used for analyzing and solving problems in communication networks

802.11g (54 Mbit/s). For the OpenFlow scenario, the data network is based on a 802.11s WMN, where the mesh interfaces are controlled by OpenFlow. We did not get results for OpenFlow at three and four hops because Iperf did not show the report. The reason is the default behavior of the OpenFlow switch, it cannot send a packet through the incoming port. However, in a WMN this behavior is valid in a node acting as a relay. This behavior caused a large amount of errors and Iperf did not show a report. As a matter of fact, batman-adv is able to achieve higher throughput at three hops, but at the expense of a higher percentage of packet loss and jitter. Finally, at four hops the throughput of batman-adv is again greater than the throughput of 802.11s, and OpenFlow was not able to pass traffic correctly. Batman-adv has an overall greater throughput, albeit at three hops 802.11s shows a better performance. The control signaling of OpenFlow have a low impact in the performance of the WMN routing protocol.

In the *jitter* measurements we found that for all hops Batman-adv had a greater jitter than the 802.11s standard, however the maximum value is still below the 20 ms permitted for a good VoIP call. The results for three and four hops are missed because the same reason of the UDP throughput test. The jitter for one hop was greater than the jitter for two hops, for all the three protocols. This is because at one hop the test was conducted using a PC for the Iperf server and one of the wireless routers as the Iperf client, which have a lower computing power.

With respect to the *loss*, it was determined that batman-adv has a higher percentage of lost datagrams than 802.11s until the third hop, while sending UDP traffic. At the fourth hop, 802.11s had too many errors and duplicated packets, so the results reported by Iperf were unreliable. The same behavior was observed for OpenFlow at three and four hops. We made tests sending *TCP traffic* to measure the maximum throughput allowed. Batman-adv obtained greater throughput than the 802.11s standard and OpenFlow. Since TCP has mechanisms for detecting and correcting errors, the throughput of batman-adv at three hops is greater than the throughput of 802.11s and OpenFlow in this case, regarding the results obtained in UDP. Besides, for all the cases batman-adv showed the best performance of the three protocols under study.

4 Conclusion

From the results of this experience we can conclude that both layer 2 routing protocols implemented with Open Source Software for WMNs have advantages and disadvantages. The B.A.T.M.A.N. Advanced protocol achieves higher transmission rates than 802.11s, but at the expense of a higher percentage of datagram loss. However, the throughput of the WMN is not an impediment for services such as videoconference; the current video codecs allow high quality videos with lower bandwidth requirements. Besides, the 802.11s showed a lower jitter than batman-adv, which is better for real-time communications. 802.11s is an IEEE standard, consequently many equipments in the future will support this protocol. Also, 802.11s is more secure because it does not have a SSID field in the frame, so it cannot be easily sniffed. Finally, 802.11s has support for multicast inherently. Regarding OpenFlow, the architecture based on a control network separated from the data network (as proposed by Dely et al. in [4]) shows an acceptable performance compared to the 802.11s standard. The in-band control approach is not recommended for WMN deployments for rural communities, due to its bad performance.

References

[1] Akyildiz, I.F.: A survey on wireless mesh networks. IEEE Communications Magazine 43(9), S23–S30 (2005)
[2] IEEE Standard for Information Technology–Telecommunications and information exchange between systems–Local and metropolitan area networks–Specific requirements Part 11: Wireless LAN Medium Access Control (MAC) and Physical Layer (PHY) specifications Am. IEEE Std 802.11s-2011, pp. 1–372 (2011)
[3] Seither, D., Konig, A., Hollick, M.: Routing performance of Wireless Mesh Networks: A practical evaluation of BATMAN advanced. In: 2011 IEEE 36th Conference on Local Computer Networks (LCN), pp. 897–904 (2011)
[4] Dely, P., Kassler, A., Bayer, N.: OpenFlow for Wireless Mesh Networks. In: 2011 Proceedings of 20th International Conference on Computer Communications and Networks (ICCCN), pp. 1–6 (2011)
[5] OpenWrt, https://openwrt.org/

Use of Open Software Tools for Data Offloading Techniques Analysis on Mobile Networks

José M. Koo, Juan P. Espino, Iván Armuelles, and Rubén Villarreal

Research Center for Information and Communication Technologies,
University of Panama, Republic of Panama
{jkoo,jp.espino,ivan.armuelles,rvillarreal}@citicup.org

Abstract. This research aims to highlight the benefits of using free software based tools for studying a LTE mobile network with realistic parameters. We will overload this LTE network and offload it through data offloading techniques such as small cells and WiFi offload. For this research, discrete-event open software network simulator ns3 will be implemented. Ns3 is a network simulator based on the programming language C++, and has all the necessary libraries to simulate an LTE and WiFi network.

Keywords: Data Offloading, WiFi, LTE, small cells, ns3, OSS for research and education.

1 Introduction

In the past few years, demand on data transfer by mobile users has rapidly increased. Among many reasons, we can mention the rapid development of technologies which have enabled users to access higher transfer speed which in turn, enable them, for example, to make better quality video calls, download high quality videos, upload more files and such. On the other hand, social networks have boosted the total amount of data which traverse mobile networks. Because of this, mobile networks undergo a cicle of ever-increasing data rates to support the growing demand of users. This is why the capacity of mobile networks is slowly reaching its limit, and this simply leads to a degradation of the quality of customer service. According to [1], by 2018, 15.9EB (1EB = 10^{18} bytes) of monthly traffic is expected to be generated by mobile devices. There are a few techniques which allow mobile service providers to increase their mobile network capacity, nonetheless, the most efficient one is frequency reuse, which is possible by reducing the size of the cell.

Among those techniques, there is "data offloading" which seeks a way to offload those users who are in the mobile network and relocate them to another network, in this case, the Internet. This way, both, those who get to stay in the mobile network and those who have been relocated, will perceive an improvement in their quality of service.

Among the variants of data offloading techniques, we have small cells and WiFi offload. Small cells are low-powered base stations mainly designed for Small-Office-Home-Office (SOHO) use. Small cells are compatible with current mobile communication technologies such as 4G and backward compatible with 2G and 3G.

On the other hand, we have WiFi offload, which is based on wireless Access point routers, through which the mobile user can connect and web-surf, make calls, among others. WiFi is considered a small cell as well, however, differ from them by the fact that WiFi operates in the unlicensed frequency spectrum.

It is necessary to evaluate the offload capacity that both techniques may offer, however, evaluation by hardware or non-open source software might be unviable, so the use of free open software tools is essential in our work.

2 Objectives

The objectives of our work are:
- Conduct a study with realistic parameters of a LTE network.
- Demonstrate that small cells and WiFi offload Access Points as data offloading techniques, reduce the load of a congested macrocell and compare the efficiency of small cells and WiFi Access Points and analyze their behaviour in tandem.

3 Methodology

To make this Project, the discrete-event open software network simulator ns3 will be used. Ns3 runs in a Linux environment and has both, LTE and WiFi libraries, which are necessary to simulate the proposed mobile network. Ns3 LTE libraries were developed by the Technical Telecommunication Center of Catalunya (CTTC) under the LENA Project. Our Project will consist of four (4) simulation scenarios which are summarized in Fig. 1.

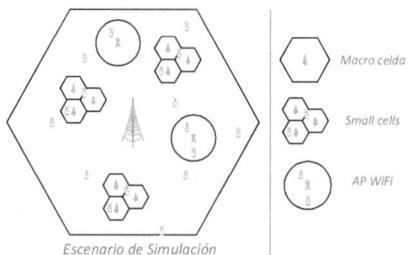

Fig. 1. Simulation scenarios and its components

References

[1] Cisco Visual Networking Index: Global Mobile Data Traffic Forecast Update, 2013-2018, Cisco
[2] SmallCellForum, http://www.smallcellforum.org
[3] ns3, http://www.nsnam.org

Crafting a Systematic Literature Review on Open-Source Platforms

Jose Teixeira and Abayomi Baiyere

TUCS - Turku Centre for Computer Science,
University of Turku,
Finland
{jose.teixeira,abayomi.baiyere}@utu.fi

Abstract. This working paper unveils the crafting of a systematic literature review on open-source platforms. The high-competitive mobile devices market, where several players such as Apple, Google, Nokia and Microsoft run a platforms-war with constant shifts in their technological strategies, is gaining increasing attention from scholars. It matters, then, to review previous literature on past platforms-wars, such as the ones from the PC and game-console industries, and assess its implications to the current mobile devices platforms-war. The paper starts by justifying the purpose and rationale behind this literature review on open-source platforms. The concepts of open-source software and computer-based platforms were then discussed both individually and in unison, in order to clarify the core-concept of "open-source platform" that guides this literature review. The detailed design of the employed methodological strategy is then presented as the central part of this paper. The paper concludes with preliminary findings organizing previous literature on open-source platforms for the purpose of guiding future research in this area.

Keywords: Open-source, FLOSS, Platforms, Ecosystems, R&D Management.

1 Introduction

1.1 Purpose and Rationale

The mobile devices market has been extremely competitive within the last five years. Apple, Google, Nokia and Microsoft among others played a very dynamic platforms-war, seeking control over the distribution of software and content to mobile hardware devices such as smartphones, netbooks and computer-tablets. The open-source software plays an important role in this platforms-war. As an indication - Apple reveals that open-source is a key part of its ongoing software strategy [1] and Google claims to lead the development of the Android platform by open-source approach [2]. On other hand, Nokia decided to give-up open-source software by closing down Symbian and Meego [3] and adopting Microsoft Windows Phone for its smart-phone strategy [4]. Yet another player; Hewlett-Packard, made big shifts on its technological strategy by abandoning WebOS, a mobile platform also based in open-source software components, after investing millions on its development[5].

An increasing number of researchers within the Information Systems (IS) field have addressed the ongoing mobile platforms-war from multiple perspectives. Mian et al. reported some implications of the open-source phenomenon on the ongoing platforms-war by studying the technological strategies employed by Apple, Google and Nokia [6]. From an innovation studies perspective, Eaton et al. explored the paradoxical relationship between control and generativity of innovation in digital ecosystem by having Apple and Google as units of analysis [7]. Building on th boundary objects theory and innovations networks literature, Ghazawneh and Henfridsson developed a process perspective of third-party development governance through boundary resources by studying the Apple's iPhone developer program [8]. From software architecture and licensing perspectives, Anvaari and Jansen evaluated the architectural openness of five different mobile platforms concluding that Google's Android and Nokia's Symbian were the most open platforms [9].

Evidently, the mobile platforms-war is gaining attention from the IS research community. However behind the *vogue,* it is important to assess how this emergent mobile platforms-war is different from previous platforms-wars covered by previous decades of published literature. This raises the following questions: Is the literature from previous platforms-wars, such as in PC and the game-console industries addressing this current war between Apple, Android, Microsoft and others?

For addressing this and other questions, we decided to execute a systematic literature review on open-source platforms, embracing a need for more and better documented literature reviews on the IS field[10]. This paper addresses the call from von Brocke et al. for the publication of two versions of the same literature review [11]. One that contains all the major findings, to be published later; and another that outlines the literature search process. This current paper addresses the latter. Subsequently, we discuss the concepts of open-source software and computer-based platforms followed by the employed methodology based on established guidelines on how to conduct a systematic literature review in the IS field.

1.2 On the Evolving Open-Source Phenomenon

There is a consensus of four freedoms expressed by Stallman [12] which laid the foundation of the open-source phenomenon [12]:

- *The freedom to run the program, for any purpose.*
- *The freedom to study how the program works and change it so it does your computing as you wish.*
- *The freedom to redistribute copies so you can help your neighbor.*
- *The freedom to distribute copies of your modified versions to others.*

For the good of the open-source community, the Open Source Initiative (OSI) was founded by Bruce Perens and Eric Raymond in 1998 to develop and maintain a more commonly agreed open-source definition, based on the social contract from the Debian Linux distribution [13]. Moreover, the OSI open-source definition introduced a novel connection between open-source software and standards [14].

According to Perens, open-source concerns not only software source code but also the distribution terms of software, as visible in the previous FSF and OSI free and open-source software definitions [15]. Both Stallman's and OSI definitions address very well the public with both expertise in software development and software license agreements, however general public could reveal difficulties in understanding the open-source term.

To position the open-source software concept used in this review with a mapping of Stallman's and OSI definitions, we propose three open-source criteria. First, the blue-print availability (software source-code is available upon request); Second, explicit intellectual propriety licenses not restricting users free-software freedoms (Software license empowers user rights); and thirdly, the compliance with standards (the software privileges the use of standards that enable interoperability).

1.3 On Computer-Based Platforms

The platform term is conceptually abstract and is widely used across many fields. Within this research, the platform term maps the concept of computer-based platform as previous addressed by Morris, Ferguson, Bresnahan, Greenstein and West [16]–[18]. As argued by West [18], platform consists of an architecture of related standards, controlled by one or more sponsoring firms [18]. The architectural standards typically encompass a processor, operating system (OS), associated peripherals, middleware, applications, etc. Platforms can be seen as systems of technologies that combine core components with complementary products and services habitually made by a variety of firms (complementors). Jointly the platform leader and its complementors form an "ecosystem" for innovation, that increases platform's value and it consequent users' adoption [19]

For example, the once leading Japanese video games industry, operate by developing the hardware consoles and its peripherals while providing a programmable software platform that allows others to develop games on top of their systems. The attraction of more game developers to the platform means more games and an increase of value for the final users (video game players). High-tech firms competing in a high-networked economy must adopt platform based strategies versus product based strategies, due to the difficulty of satisfying an increasing complex consumer demand [20]. In the development of certain complex systems, an "all in house" strategy might not be economically feasible, organizations must adopt "platform-thinking" and focus efforts on the highest value-adding components of the platform, making it open and attractive to all possible participants.

Within this research, the authors address literature on open-source computer-based platforms: meaning computer-based platforms that not only integrate open-source software components, but also provide a set of publicly available open-source components. Prominent examples can be the Google's Android, Apple iOS and Nokia Maemo platforms empowering mobile devices. For instance, all the vendors integrate the WebKit open-source web browser engine into their platforms while providing their modified WebKit versions in open-source manners. It is important to note that, within this reviews context, computer-based platforms combine hardware and software but can also be pure-software platforms. Krishnamurthy and Tripathi [22] and Teixeira [23] studied platforms structured over pure software artefacts. Platforms leaders provide and set the boundaries of their technological-core and provide additional development mechanisms that allow third-parties to complement while adding value to the overall platform under network effects.

2 Research Methodology and Design

After clarifying the core-concepts of "open-source software" and "computer-based platforms" we present the methodology used for the review in this section.

2.1 Research Goals and Methodological Base

Primarily and most importantly, by conducting this structured literature review the authors aim to provide an aggregated vision of what is well known within the academia regarding open-source platforms. The underlying research questions are:

- **RQ1:** What are the seminal works bridging open source and platforms?
- **RQ2:** Does the literature from previous platforms-wars, such as in the PC and the game-console industries, address this current mobile-platforms war between Apple, Android, Microsoft and others?
- **RQ3:** What is the seminal literature to be taken into account by researchers and practitioners addressing the ongoing mobile platforms-war?
- **RQ4:** Which previous research findings can't be generalized for such novel and contemporary scenario?

This review considers methodological guidelines provided by Webster and Watson [10], Järvinen [23], von Brocke et. al. [11] and Okoli and Schabram [24]. Transparency and rigour in documenting the literature review process, the use of a systematic and future reproducible procedure; were some of the base-pillars of this review. Simple and common available software tools, like spreadsheet software (LibreOffice), citation manager (Zotero), graph visualization software (Graphviz) and a mind-mapping tool (Xmind) eased the literature review process.

The literature review process started in November 2010, the final set of articles were retrieved on March 2011 and were carefully read and analyzed while taking in account the different methodological guidelines on conducting a literature review.

2.2 Design and Research Basis

After reading the literature review guides and analyzing a small set of systematic review articles published in the IS field, the authors decided to follow closely the literature review design from von Brocke and Theresa [25]. As in [25], the authors made use of Emerald, EBSCO and ProQuest ABI/Inform databases of general journals and conferences; plus the use of Google books index on published books;

and finally the use of the eLibrary system from the Association of Information Systems (AIS) as a database indexing more specific journals and conferences within IS field.

The authors decided to complement von Brocke and Theresa research basis by including the Volter national database that indexes books within a large national libraries network. The following Table 1 summaries the database sources from where the literature was collected. All databases were accessed using authors host University Internet proxy even if accessed remotely.

Table 1. Database sources used for conducting the literature review

Source	Type	Website http://
Emerald	General journals and conferences	emeraldinsight.com
EBSCO	General journals and conferences	web.ebscohost.com
ProQuest ABI/ Inform	General journals and conferences	search.proquest.com
Google books	Published books	books.google.com
Volter database	Published books	volter.linneanet.fi
AIS eLibrary	IS journals and conferences	aisel.aisnet.org

For retrieving literature that aggregates both knowledge on open-source and computer-based platforms, previous knowledge of the authors reinforced by discussions within the academic circle influenced the choice of the keywords for the search. Addressing the open-source term, the keywords "open source", "open-source", "OSS", "FLOSS" and "libre" were used. Moreover, for capturing relevant literature within computer-based platforms the keywords "platform", "platforms", "platform-based", "eco system", "eco systems", "eco-system" and "eco-systems" were employed. The decision to use several keywords increased the amount of relevant literature included in the review.

The authors discarded publications that were not relevant to the IS field after a careful content analysis. Most of the publications discarded did not fit with this papers adopted definitions of open-source and platforms. The authors documented and tabulated each discarded publication item while building associated exclusion criteria. The search was limited to peer-reviewed publications; several published books and journal articles were discarded because they did not clearly meet this criterion. The search was also limited to research expressed in the English-language.

The research basis (α) was defined by searching, within the mentioned source databases, for publication items with both open-source and platforms keywords on their titles. An initial set of fifteen publications were defined as the starting point for our research. The fifteen publications included books, two conference proceedings and the remaining were serial journals. After an extensive analysis of the research basis, the authors decided to extend the research (β) by searching for articles with keywords capturing open-source on their titles and with keywords capturing platforms on the abstract. A total of 360 new publications were identified with this first research extension. For future research, the authors consider the possibility of extending the research to include other publications with platforms on the title and open-source on the abstract.

The following Table 2 gives an overview on how the captured research publications were retrieved by each of the six source databases. A considerable number of collisions, publications indexed more by different databases, was encountered. Books and dissertation databases were not considered in our research extension because books databases do not support queries addressing a possible book abstract.

Table 2. Number of captured research publications per database source

	EME	EBS	PQA	GOB	VOL	IAS
α (title CONTAINS (open-source AND platforms))	0	3	5	6	0	1
β ((title CONTAINS open-source) AND (abstract CONTAINS platforms)	1	24	318	QNS	QNS	2
Total captured items per database source	1	27	323	6	0	3

2.3 Extraction and Categorization of Literature

In order to provide both a quantitative and qualitative overview of relevant research of open-source platforms within the Informations Systems field, the authors extracted and categorized the literatures according to their meta-description and content. The authors first conducted a simpler categorization of the literature without looking at its full content. Some of the retrieved articles were discarded by its meta-description (i.e. after reading the abstract). After reading each articles meta-description, the authors moved afterwards to a more demanding phase, where the deep reading of each papers content enabled the extraction and categorization of research on open-source platforms.

For the first step, content independent information was extracted and categorized by using meta-descriptions of each captured paper. Not all non-content information was available within the used sources databases, requiring visits to the different publishers Internet resources. For each paper found, a manual citation analysis was made using both the http://scholar.google.com and the http://www.isiknowledge.com web resources. The authors decided to keep track of each captured research paper price, if applicable: both the payment amount charged by the publisher to download the paper and the yearly subscription rate, all for later arguing on the cost of this literature review.

For the second step, and in order to provide a qualitative overview of the literature review, the authors delved into the articles content. This started an ongoing demanding analysis of each captured paper, identifying key information such as research questions, methodology, outlined future research, research propositions, theoretical implications, implications for practice, key references, among other content information. After full paper reading and using spreadsheets, the authors systematically retrieved for each paper, information about the research questions being addressed; their triggers and motivations, implications for theory and practice, methodology and philosophical standings, perceived theoretical and empirical relevance, etc. For the very specific context of this literature review the authors also captured for each item what are the research related industry verticals and platforms being studied. For each paper, the authors complemented the collected information in a spreadsheet with two to three slides containing the message of each paper,

After the content analysis, the authors made the transition from author to concept-centric approach as suggested by Webster and Watson [10]. A long concept matrix was developed for mapping the analyzed publication items with key concepts that emerged during the literature review process, e.g. the concepts of "Community of Practice" [26] and "Sense of Community" [27]. Relationships between these key concepts were then mapped using diagram tools (i.e Graphviz and Xmind) providing a theoretical overview[1] of previous research in open-source platforms.

3 Preliminary Findings

As previously mentioned, this literature review is still a work in progress. So far, the meta-description analysis of the retrieved 360 articles is completed. However; just 170 of the articles has been fully read and content-analyzed. This literature review is aimed to be systematic, rigorous and exhaustive which turned out to be a slow process lasting several years. In this section, we present our preliminary findings by revisiting the initial research questions and outlining future research.

3.1 Revisiting the Research Questions

The first research question was "What are the seminal works on open-source platforms?". Based on a citation analysis of the retrieved publications on Google and Thomson Reuters services; and by its recurrence within the articles analyzed so far, the authors proposes: The economic works of Economides and Katsamakas[29]; the open-source adoption studies of Dedrick and West[14]; [30] and the R&D management strategy work of West[18]; as seminal works on open-source platforms.

Our second research question inquired "if research addressing the current mobile-platforms takes in consideration literature from previous platforms-wars?" The third and related initial research question is "What is the seminal literature to be taken in account by researchers and practitioners addressing the ongoing mobile platforms-war?" After reviewing ad-hoc emergent literature on the novel mobile-platforms war such as: Basole's visualization in a converging mobile ecosystem[31]; Eaton et al. description of the paradoxical relationship between control and generativity on Apple and Google ecosystems[7] ; or the innovation study from Remneland-Wikhamn et al. on the iPhone and Android mobile platforms; we claim that emergent research addressing the current mobile-platforms is not considering, or exploiting previous seminal works on open-source platforms, as it often should.

Out last initial research question inquired if previous research findings, on previous platforms-wars, can be generalized to the current mobile platforms-war, scenario. Previous seminal works from Economides, Katsamakas, Dedrick and West [14], [18], [29], [30] assume a scenario where open-source is an alternative strategy for low-cost

---

[1] Theoretical overview within Gregor's nature of theory in information systems research [28].

players, with reduced market-share, against more successful corporations enjoying a quasi-monopoly situation. Using researchers own words:

> *"When a system based on an open source platform with an independent proprietary application competes with a proprietary system, the proprietary system is likely to dominate the open source platform industry both in terms of market share and profitability. This may explain the dominance of Microsoft in the market for PC operating systems."* in [29]

> *"On the other hand, Microsoft's proprietary platform strategies continued to be successful"* in [18]

> *"The most important driver of adoption was cost "* in [14]

> *"The major factors are cost, perceived reliability, compatibility ..."* in [30]

Tables turned: First of all, open-source is no longer associated with low-cost products within the current mobile platforms-war. Moreover, the traditional proprietary software players, such as Microsoft and Blackberry, are currently struggling with residual sales on the mobile devices market [32]. Apple, Google and Google Android partners are effectively dominating the market, while charging more for their high-end devices than their competitors[33], all with strategies that esteem open-source software[1], [2].

3.2 Future Research

When contrasting previous literature on older "platforms-wars", such as the ones from the PC and game-console industries, with the current and under-studied mobile platforms-war, we empirically notice that many of the market players remain the same (Microsoft and Apple). There is a scenario of convergence: same firms push for similar technological standards across different platforms, i.e. Microsoft Windows within X-box, Surface Tablets, PC, Netbooks and Mobile phones. This convergence between industries remains unexplored by academia. Interesting research questions dealing with the implications of such convergence remain unexplored, i.e "should firms concentrate on one platform-war or run several platform-wars in parallel?

References

[1] Apple, "Apple | Open Source" (December 06, 2012),
http://www.apple.com/opensource/ (accessed December 06, 2012)

[2] Android, "Welcome to Android" (December 06, 2012),
http://source.android.com/ (accessed December 06, 2012)

[3] TheGuardian, "Nokia closes Symbian to the world" (November 29, 2010),
http://www.guardian.co.uk/technology (accessed December 06, 2012)

[4] Nokia "Nokia and Microsoft announce plans for a broad strategic partnership to build a new global ecosystem» Nokia – Press" (February 11, 2011),
http://press.nokia.com/2011/02/11/ (accessed December 06, 2012)

[5] HP Financial news, "HP Investor Relations" (December 06, 2012), http://h30261.www3.hp.com/ (accessed December 06, 2012)
[6] Mian, S.Q., Teixeira, J., Koskivaara, E.: Open-Source Software Implications in the Competitive Mobile Platforms Market. In: Skersys, T., Butleris, R., Nemuraite, L., Suomi, R. (eds.) I3E 2011. IFIP AICT, vol. 353, pp. 110–128. Springer, Heidelberg (2011)
[7] Eaton, B., Elaluf-Calderwood, S., Sørensen, C., Yoo, Y.: Dynamic structures of control and generativity in digital ecosystem service innovation: the cases of the Apple and Google mobile app stores. London School of Economics and Political Science (2011)
[8] Ghazawneh, A., Henfridsson, O.: Governing third-party development through platform boundary resources. In: ICIS Proceedings, pp. 1–18 (2010)
[9] Anvaari, M., Jansen, S.: Evaluating architectural openness in mobile software platforms. In: Proceedings of the Fourth European Conference on Software Architecture: Companion Volume, pp. 85–92 (2010)
[10] Webster, J., Watson, R.: Analyzing the past to prepare for the future: Writing a literature review. MIS Quarterly 26, 13–23 (2002)
[11] von Brocke, J., Simons, A., Niehaves, B., Riemer, K., Plattfaut, R., Cleven, A.: Reconstructing the Giant: On the Importance of Rigour in Documenting the Literature Search Process, Verona, pp. 2206–2217 (2009)
[12] Stallman, R.: The GNU manifesto (1985)
[13] Debian, "Debian Social Contract (version 1.1)" (1997), http://www.debian.org/social_contract#guidelines (accessed March 29, 2011)
[14] Dedrick, J., West, J.: An exploratory study into open source platform adoption. In: Proceedings of the 37th Annual Hawaii International Conference on System Sciences, p. 10 (2004)
[15] Perens, B.: The Open Source Definition. In: Open Sources: Voices from the Open Source Revolution, 1st edn., p. 280. O'Reilly Media, Inc. (1999)
[16] Ferguson, C.H., Morris, C.R.: Computer Wars: How the West Can Win in a Post-IBM World. Times Books New York, NY (1993)
[17] Bresnahan, T.F., Greenstein, S.: Technological Competition and the Structure of the Computer Industry. The Journal of Industrial Economics 47(1), 1–40 (1999)
[18] West, J.: How open is open enough?: Melding proprietary and open source platform strategies. Research Policy 32(7), 1259–1285 (2003)
[19] Gawer, A., Cusumano, M.A.: How companies become platform leaders. MIT/Sloan Management Review 49 (2008)
[20] Hagiu, A.: Japan's High-Technology Computer-Based Industries: Software Platforms Anyone? RIETI Column (October 19, 2004)
[21] Teixeira, J.: Open-Source Technologies Realizing Social Networks: A Multiple Descriptive Case-Study. In: Hammouda, I., Lundell, B., Mikkonen, T., Scacchi, W. (eds.) OSS 2012. IFIP AICT, vol. 378, pp. 250–255. Springer, Heidelberg (2012)
[22] Krishnamurthy, S., Tripathi, A.K.: Monetary donations to an open source software platform. Research Policy 38(2), 404 (2009)
[23] Järvinen, P.: On developing and evaluating of the literature review. In: The 31st Information Systems Research Seminar in Scandinavia, Workshop 3 (2008)
[24] Okoli, C., Schabram, K.: A guide to conducting a systematic literature review of information systems research (2010)
[25] von Brocke, J., Theresa, S.: Culture in Business Process Management: A Literature Review. Business Process Management Journal 17(2) (2011)

[26] Lave, J., Wenger, E.: Situated learning: Legitimate peripheral participation. Cambridge University Press (1991)
[27] Chavis, D.M., Hogge, J.H., McMillan, D.W., Wandersman, A.: Sense of community through Brunswik's lens: A first look. J. Community Psychol. 14(1), 24–40 (1986)
[28] Gregor, S.: The nature of theory in information systems. Mis Quarterly 30(3), 611–642 (2006)
[29] Economides, N., Katsamakas, E.: Two-sided competition of proprietary vs. open source technology platforms and the implications for the software industry. Management Science 52(7), 1057–1071 (2006)
[30] Dedrick, J., West, J.: Why firms adopt open source platforms: a grounded theory of innovation and standards adoption. In: Proceedings of the Workshop on Standard Making: A Critical Research Frontier for Information Systems, pp. 236–257 (2003)
[31] Basole, R.C.: Visualization of interfirm relations in a converging mobile ecosystem. Journal of Information Technology 24(2), 144–159 (2009)
[32] Gartner, "Smartphone Sales Accounted for 55 Percent of Overall Mobile Phone Sales," http://www.gartner.com/ (accessed November 21, 2013)
[33] iPhone 5S vs. Galaxy S4 And The Rest: Your Guide To Buying The Right Smartphone, http://www.huffingtonpost.com/2013/09/19/ (accessed November 21, 2013)

Considerations Regarding the Creation of a Post-graduate Master's Degree in Free Software

Sergio Raúl Montes León[1,2], Gregorio Robles[2],
Jesús M. González-Barahona[2], and Luis E. Sánchez C.[1]

[1] Research Group GSyA, Universidad de las Fuerzas Armadas (ESPE-L),
Latacunga, Ecuador
smontes@espe.edu.ec, luisenrique@sanchezcrespo.org
[2] Universidad Rey Juan Carlos, Madrid, España
{grex,jgb}@gsyc.urjc.es

Abstract. Free software has gained importance over the last few years, and can be found in almost any sphere in which 'software processes' are important. However, even when universities and higher education establishments include subjects concerning free programming and technologies in their curriculums, their graduates tend to attain limited technological, organisational and philosophical knowledge that limits them as regards their participation in, management and development of free software projects. This gap in skills and knowledge has recently led to a series of post-graduate studies whose objective is to offer students the possibility of acquiring competencies that will allow them to become experts in free software. This paper presents a study concerning the offers for post-graduate studies in free software that currently exist, with the intention of creating similar post-graduate studies in Ecuador.

1 Introduction

In the present-day world, society is developed on the basis of information technology, and software has a particularly important function in this, thus demonstrating that humanity's knowledge has evolved by means of computing. For this knowledge to be within everyone's reach, the use of free software is essential. Free software can, moreover, currently be found in a multitude of environments, if not in all of them, and has come to be of prime importance over the last few years [5].

European and South American governments have now created laws and decrees for the use of free software [1], which in Ecuador consists of Decree 1014. The need to train personnel who are qualified in the sphere of free software is therefore being investigated. The European Union has recommended that research should be carried out in this sphere, since it alleges that the habitual lack of knowledge as regards code does not permit the auditing of real functioning, which could seriously compromise the security of some countries, thus leaving them in the hands of companies that create private programmes [3].

We should also mention that free software attracts the attention of companies and public administrations throughout the world, and it is in this way that countries such as Spain, Brazil, Mexico, Venezuela, Columbia, Peru, Chile, Argentina and, essentially, Ecuador foment its use and development [4],What is more, many large technological companies such as IBM, Apple, Facebook or Google support the free software movement by both freeing some of their star products, such as WebKit, MySQL, Android, and participating in the development of projects such as Eclipse and Linux, among others [6].

Even when universities and higher education establishments include free technologies in their curriculums, their graduates normally lack the training needed to be able to carry out the tasks that are necessary to successfully participate in a free software project [2]. It is therefore common for universities to teach programming. However, although this is one vital requirement as regards providing code, it is not the only one, since to be able to take part in a free software project it is also necessary to have knowledge of development tools (e.g. the version system), conventions (such as sending code) and even organisational skills (who to ask).

The existence of this knowledge gap is the reason why university graduates must therefore supplement their knowledge by means of auto-didactic learning. This gap in curriculums has therefore led to the emergence of proposals for postgraduate studies that will allow students to acquire the skills and knowledge needed to become experts in free software. This type of studies includes not only technological aspects, but also the management of projects, business models and even philosophical aspects.

This paper is organised as follows: the methodology used in this research is described in Section 2, while Section 3 shows a study carried out in order to create a post-graduate course in free software in Ecuador. The results of the analysis of post-graduate courses offered by some possible universities are presented in Section 4, while our conclusions and future work are shown in Section 5.

2 Methodology

Much of the research that is carried out considers a literature review, although the eventual objective is not the review in itself, but rather that it constitutes a technique that can be used to understand the state-of-the-art of the theme being tackled. What is more, a state-of-the art study constitutes the basis for the formulation of proposals of greater reach.

Computing disciplines, of which software is one, are very recent in comparison to other science disciplines, signifying that there are no methodologies with which to guide the development of systematic reviews in them, Kitchenhan [7], therefore proposed a method with which to carry out systematic [10] reviews that is based on the guidelines developed for medical research which were adapted to be used by a team of researchers in the sphere of software engineering.

In this research, the systematic review has been carried out by using the Google Scholar search engine to search for references to 'Master's degrees in Free

Software'. The results and information obtained from its websites are shown in the following sections.

3 Post-Graduate Studies in Free Software

According to the Royal Academy for the Spanish Language, a Master's degree is a post-graduate course in a particular speciality, while a post-graduate course is a cycle of specialisation studies that take place after graduation.

The Higher Education Organic Law of Ecuador[1], in Art. 120, defines it as follows: "it is an academic degree which seeks to broaden, develop and explore in greater depth a discipline or specific area of knowledge. It provides people with the tools that will enable them to explore the field of knowledge in greater depth, both theoretically and instrumentally".

3.1 Openings for Professionals with Master's Degrees in Free Software

A Master's degree in Free Software provides professionals with capacities in the four areas defined in the professional capacities profile report generic to ICT created by the *Career-Space*[2] [9] consortium that are necessary to carry out their activities as regards all that is related to the use, application and development of In each of the areas, the qualification re-enforces the following professional roles:

- Technicians in communications software development.
- Software project managers.
- Communication network designers.
- Application programmers.
- Software engineers.
- Information technology business consultants.
- Electonics business consultants.
- Business analysts.
- Information strategy management consultants.
- Systems implementation technicians.
- Integration systems technicians.
- Project managers ITC.

---

[1] LOES by its acronym in Spanish: http://www.utelvt.edu.ec/LOES_2010.pdf
[2] Career Space is a consortium that is formed of eleven ICT companies and analyses the need to provide professionals with capacities in this sphere
http://www.space-careers.com/

3.2 Existing Free Software Master's Degrees

The Internet has been used to search for academic offers of Master's degrees in Free Software in Europe and South America [8]. It has thus been possible to find the following university entities, which offer the following studies[3]:

- Official Master's degree in Free Software from the Universidad Rey Juan Carlos[4] (Madrid, Spain),
- Official Master's degree in Free Software from the Universitat Oberta de Catalunya (Open University of Catalonia)[5] (UOC) (Catalonia, Spain),
- Master's degree in Open Code Software from the University Institute of Lisbon[6] (ISCTEC, Lisbon, Portugal),
- Master's degree in Free Software Engineering (MISWL) from the Polytechnic Senior Lleida[7] (Lleida, Spain),
- Master's degree in Free Software from the Autonomous University of Bucaramanga[8] (UNAB, Bucaramanga, Colombia),
- Master's degree in Free Software from the Autonomous University of Chihuahua[9] (UNACHI, Chihuahua, Mexico),
- Master's degree in Free Software from the University of Extremadura[10] (Extremadura, Spain),
- and the Master's degree in Software Open Source Management from the University of Pisa[11] (Pisa, Italy).

3.3 Analysis of Existing Master's Degrees in Free Software

Having identified the Master's degrees in Free Software, we shall now go on to investigate each one in detail using the publicly provided information from their websites. Table 1 provides a summary of the most important qualititive aspects of these Master's degrees, which are classified in the following manner:

- Country: This refers to the country in which the Master's degree is offered.
- Modality: This refers to the mode of studies, such as (Physical presence, Virtual).

---

[3] Please note that this study is not intended to be complete, but is rather a representative sample of Master's degrees in Free Software. The authors would like to apologise for any other initiative that has been omitted from this study.
[4] http://master.libresoft.es. This Master's degree is also offered in conjunction with the Igalia company in Galicia
[5] http://estudios.uoc.edu/es/masters-universitarios/software-libre/
[6] http://iscte-iul.pt/cursos/mestrados/10702/apresentacao.aspx
[7] http://www.udl.cat/estudis/masters_cast/programario_libre.html
[8] http://www.unab.edu.co/portal/page/portal/UNAB/programas-academicos/software-libre-virtual?programa=MSOL
[9] http://www.uach.mx/investigacion_y_posgrado/2010/10/28/maestria_en_software_libre/
[10] http://www.eweb.unex.es/eweb/msl/index.html
[11] http://www.master.netseven.it/index.php?page=/master/home

- Duration: Length of time that the Master's degree course lasts.
- Professionals: Towards which type of professionals the Master's degree is oriented, such as: Information and Communication Technologies (ICT), General (any professional with a university degree).
- Credits: This refers to the ECTS.
- Placements: This refers to whether or not the student is required to participate in an industrial placement during the Master's degree.
- Final Project: This refers to whether or not the student is required to produce an end-of-degree project.
- Options: This refers to whether or not the Master's degree contains optional subjects.
- Agreement: This refers to whether or not the Master's degree is carried out via agreements with other universities.
- NK: This refers to the fact that Nothing is Known about the aspect in question, and no information is provided on the website.

Table 1. Comparison of Universities Offering Master's degrees in Free Software

Comparison of Universities Offering Master's degrees in Free Software								
	Universities							
	URJC	UOC	ISTEC	U Lleida	UNAB	UNACHI	Extremadura	U Pisa
Country	Spain	Spain	Portugal	Spain	Colombia	Mexico	Spain	Italy
modality	Presence	Virtual	Presence	Presence	Virtual	Presence	Presence	Presence
Duration	1 year	1 year	1 year	2 year	2 year	NK	1 year	1 year
Professional	General	General	ICT	ICT	ICT	ICT	ICT	ICT
Credits	60	60	60	60	43	NK	60	60
Practices	yes	no	no	yes	no	Nk	Nk	Nk
Final Project	yes	yes	no	yes	yes	Nk	Nk	Nk
Electives	yes	yes	no	no	no	Nk	Nk	Nk
Agreement	UOC	no	UOC	no	UOC	Nk	Nk	Nk

The students for whom the Master's degrees are intended are those with a technical professional profile (ICT), since the objective of these degrees is to study technology and skills in greater depth, thus supposing that the students will already be knowledgeable as regards programming and engineering. Two exceptional cases are those of the Master's degrees offered by the URJC[12] and the UOC, since these universities have broadened their curriculums to include any type of entrance profile and offer, on the one hand level 0 subjects in order to allow students to attain the previous requirements for the subjects, and on the other a wide range of subjects including those with a socio-technical element, such as communication management, business models, etc

There are Master's degrees which are taught both in the students' physical presence or virtually, although in the cases in which the students are actually present, we have observed the use of an e-learning environment (generally Moodle[13]), which has led us to conclude that this type of teaching is, to a great extent,

---
[12] http://master.libresoft.es
[13] http://www.moodle.org

backed up with on-line elements. The URJC explicitly states that its Master's degree is considered to be a form of *blended learning*, since many of the course activities are carried out virtually, and the students' physical presence is limited to that required by law.

The majority of the Master's degrees are designed to last one or two years, although this is initially delimited by the students' dedication. This signifies that the most interesting column in the table is that which indicates the ECTS[14] credits that the student must exceed to obtain the degree. One ECTS credit is obtained after an average of 25 hours' work by each student, including classes in which they are physically present, the preparation needed for these classes, the tasks and activities carried out, exam preparation, and even the exams themselves. It will thus be observed that, in general, Master's degree students must exceed 60 ECTS credits, which is equivalent to a year's work by the student. The academic years may therefore be longer or shorter, depending on the amount of time available to the student. The universities offer various possibilities in order to adapt to the students. The Master's degree at the URJC can therefore be obtained by solely attending classes on Fridays – rather than Thursdays and Fridays – throughout a single academic year. The subjects that take place on Thursdays in even academic years are moved to Fridays in the odd academic years, and vice versa. This facilitates the students' participation, since they are from professional environments, signifying that their time is always limited. Another example is the UOC which allows its students, who are generally professionals with families, to take the course at their own pace, choosing 10 or 15 ECTS per course.

The design of the Master's course curriculums may differ as regards certain characteristics. We have therefore investigated three of these elements:

- The need to go on industrial placements. On some Master's courses, the students must show that they have had experience in professional environments related to free software. They are therefore required to go on industrial placements at companies in the sector at which they gain experience. At some universities agreements have not only been reached with software companies, but also with foundations, thus allowing the students to carry out the practical element of the course in foundation projects.
- The realisation of an end-of-degree project. In order to obtain the degree it is compulsory to produce an end-of course work which is, if possible, a research work that is carried out by the student under the supervision of a tutor. This work must be presented in the form of a report and must be publicly defended in front of a committee.
- Options: The Master's degree curriculums may offer a series of subjects from which the students may choose. These options allow the student to specialise, and are therefore highly advantageous. However, their drawback is that they only make sense if there are a considerable amount of students enrolled on the course.

---

[14] ECTS (European Credit Transfer System) (BOE 2003). ECTS credits are established by measuring the amount of work that the student has carried out both inside and outside the classroom to pass a subject.

Table 2. Curricular Planning of Free Software Master's Degree (URJC)

Name of subject	Description of Minimum Content	Type	ECTS	Seminars	Character
Introduction to free software	Introduction, motivations, definition and history of free software. Introduction to legal, economic, social and technological framework of free software	theorist	3	1º	obligatory
Legal aspects of free software	Intellectual property, legal aspects, licences	theorist	3	1º	obligatory
Economic aspects of free software	Introduction to economic aspects of software, types of business models, business plans and case studies.	theorist	3	1º	obligatory
Developers and their motivations	Profile of developers, motivations, roles and leadership in free software. Evolution of participation and integration programmes in projects	theorist-practical	3	1º	obligatory
Free software development. Tools	Development environments, version control systems, defect and task management in free software projects. IDEs and collaborative tools. Case studies specific to free software project development	practical	3	1º	obligatory
Free software project evaluation	Introduction to software quality. Light evaluation methodologies. Project information extraction tools, quality evaluation atomisation and specific case studies	practical	3	1º	obligatory
Case studies I - II	Free projects, including themes related to technological, organisational, legal, economic and governmental themes	seminary	9	1º y 2º	obligatory
Free software Project management	Introduction to free software Project creation, Infrastructure, communication and management of community elements	theorist	3	2º	obligatory
Advanced free software development	Advanced tools and advanced technical aspects of free software development.	practical	3	2º	elective
System integration	Introduction to systems administration, storage, networks, security and virtualisation with free software	practical	3	2º	elective
Free software instalment	Instalment of free software. Free software in desktops and servers. Cost analyses, requirements and study of infrastructure deployment using free software	theorist	3	2º	elective
Free software communities	Free software communities from the empirical viewpoint, data collection tools, database management. Introduction to the evolution of software and the study of free software communities. Specific case studies	practical	3	2º	elective
Industrial placements	Industrial placements at companies	practical	12	1º y 2º	obligatory
End-of-degree project	Development of Project focused on free software	Project	12	1º y 2º	obligatory

Finally, we have observed that many universities have recognised agreements with the Open University of Catalonia (UOC). This allows students to begin their studies by being physically present and then, if necessary, change to the UOC, which is completely on-line.

3.4 Curricular Design

The Master's degree curriculums are organised in a series of subjects. It is interesting to note, as will be shown below, that there is a great similarity in the Master's degrees studied as regards the contents that are offered. This is because there have been many cases of close collaboration among the universities that offer a Master's degree in free software when designing their curriculums. The Master's degree that has been in existence for the longest period of time, which is that offered by the Open University of Catalonia since 2003, is therefore assessed by professors from the URJC. Since this was the first Master's degree on this theme, and owing to other universities' agreements with the UOC, it is possible to find this general schema in other Master's degrees.

This paper shows the curricular design of the Master's degree offered by the URJC, which is divided into subjects of 3 ECTS (75 hours' worth of dedication per student), while that of the UOC has subjects of 5 ECTS (125 hours per student). This signifies that the Master's degree from the URJC, although it has the same content as that of the UOC, is atomised to a greater degree, thus permitting us to show the contents on offer in a better manner.

Table 2 shows the subjects in the Master's degree at the Universidad Rey Juan Carlos, together with a description of contents, the amount of ECTS credits, the semester in which they are taught, and whether the subjects are obligatory or optional. In order for this degree to be awarded the title of University Master's degree (the official title of the European Higher Education Area – EHEA – and therefore valid in all the countries in the OECD), its course design had to be approved by the Spanish ANECA agency (National Agency for the Evaluation of Quality and Accreditation).

According to their type, the subjects can be classified as: (i) theoretical: subjects taught with the purpose of the students acquiring knowledge; (ii) practical: subjects which are focused on the acquisition of technological skills, with a highly practical structure; (iii) seminaries: A cycle of talks, generally given by speakers invited from the free software community, at which the students can gain first-hand experience of the *reality* of the movement; and (iv) ohers: The practicum and final work of the Master's degree, explained previously.

The Master's degree contains four optional subjects, of which two must be chosen. Two itineraries are therefore recommended: (a) *Technological*, which consists of the subjects *Systems integration* and *Advanced free software development*, and is intended for those students who wish to develop a more technological profile; and (b) *Management*, which consists of the subjects *Free software installation* and *Free software communities*, and is intended for those students who prefer to specialise in community management and consultation tasks at an organisational and deployment level.

4 Analysis and Results

The analysis carried out has allowed us to conclude that, as is shown in Table 3 the Master's degrees in free software offered by other universities share a schema similar to that of the Universidad Rey Juan Carlos.

As will be noted, there are various common contents, although some subjects are only offered by the URJC and are of two types: (a) The *Case Studies I* and *Case Studies II* eminaries which, owing to their idiosyncratic nature, only take place at the URJC, and (b) Subjects with which the URJC has research experience, as is the case of *Free software Project evaluation* or *Developers and their motivations*. In order to cover these subjects, the other Master's degrees are inclined to offer technological subjects related to GNU/Linux systems administration, web application development or databases. Table 3 shows a comparison of the universities that currently offer a Master's degree in free software with regard to the URJC's schema of subjects.

Table 3. Analysis of Universities' Subjects with Regard to the URJC

Subject	URJC	UOC	ISTEC	U LLEIDA	UNAB	UNACHI
Introduction to free software	yes	yes	yes	yes	yes	NK
Legal aspects of free software	yes	yes	yes	yes	yes	NK
Economic aspects of free software	yes	yes	yes	yes	yes	NK
Developers and their motivations	yes	no	no	no	no	NK
Free software development. Tools	yes	yes	yes	yes	yes	NK
Free software project evaluation	yes	no	no	no	no	NK
Case studies I	yes	no	yes	yes	yes	NK
Free software Project management	yes	yes	no	yes	no	NK
Case studies II	yes	no	yes	yes	yes	NK
Prácticum	yes	yes	yes	yes	no	NK
End-of-degree project	yes	yes	yes	yes	yes	NK
Advanced free software development	yes	yes	yes	yes	yes	NK
System integration	yes	yes	no	yes	yes	NK
Free software instalment	yes	yes	yes	yes	yes	NK
Free software communities	yes	no	no	no	no	NK

5 Conclusions and Future Work

This paper presents a study of the offers that are currently available as regards post-graduate studies in the field of free software. It begins by arguing why it is important to have experts in free software at present, and by explaining that students finish their university studies without having attained the knowledge and skills needed by the free software industry and community. Universities and higher education establishments therefore have the possibility of teaching these aspects by offering post-graduate degrees that are oriented towards those professional profiles that wish to specialise in the area of free software [8].

Various European and South American post-graduate initiatives have been found, and the degree of affinity as regards the content of their curriculums is shown in Table 3, principally with regard to historic and organisational matters. The curricular schema of the Universidad Rey Juan Carlos has therefore been used, which provides details of: the existence of a general curriculum, along with the materials available that facilitate the tasks of those universities that wish to offer a Master's degree in free software.

As will be observed in the section in Table 1 entitled 'Analysis of existing Master's degrees in free software', the research has not located any Master's degrees of this kind in English-speaking countries.

This research will continue to review whether professional university graduates who have taken post-graduate degrees in free software have more possibilities of obtaining employment. We shall also verify the evolution that the universities offering Master's degrees in free software have undergone.

Acknowledgements. This work has been funded by the Spanish Government with the SobreSale project (TIN2011-28110), by the Comunidad de Madrid with the Red de Excelencia eMadrid (S2009/TIC-1650), by the Prometeo Project of the Ministry of Higher Education Science, Technology and Innovation of the

Republic of Ecuador, and by the SIGMA-CC project (Ministerio de Economía y Competitividad and Fondo Europeo de Desarrollo Regional FEDER, TIN2012-36904).

References

1. Alves, A., Stefanuto, G., Castro, P., Pessôa, M.: Brazilian public software and quality, pp. 413–415 (2010)
2. Bouckaert, R.R., Frank, E., Hall, M.A., Holmes, G., Pfahringer, B., Reutemann, P., Witten, I.H.: Weka—experiences with a java open-source project. J. Mach. Learn. Res. 11, 2533–2541 (2010)
3. Bouras, C., Filopoulos, A., Kokkinos, V., Michalopoulos, S., Papadopoulos, D., Tseliou, G.: Guidelines for the procurement of free and open source software in public administrations, pp. 29–36 (2012), cited By (since 1996)
4. Chan, A.: Coding free software, coding free states: Free software legislation and the politics of code in peru. Anthropological Quarterly 77(3), 531–545 (2004)
5. Haick, B., Klautau, A.: Free software tools for IT management and processes organization, vol. 1, pp. 219–223 (2013)
6. Honda, M., Kobayashi, M., Nagumo, M., Kawakatsu, Y.: Android software platform development at fujitsu. Fujitsu Scient. and Tech. J. 49(2), 238–244 (2013)
7. Kitchenham, B., Pretorius, R., Budgen, D., Brereton, O.P., Turner, M., Niazi, M., Linkman, S.: Systematic literature reviews in software engineering - a tertiary study. Information and Software Technology 52(8), 792–805 (2010)
8. Montes-León, S.R.: Propuesta para la creación de un máster en software libre en la escuela politécnica del ejército extensin latacunga en ecuador. Trabajo fin de máster, Universidad Rey Juan Carlos, España (Septiembre 2012)
9. Space, C.: Perfiles de capacidades profesionales genéricas de las tic. Centro Europeo para el desarrollo de la formación profesional (Junio 2001)
10. von Wangenheim, C.G., Hauck, J.C.R., Salviano, C.F., von Wangenheim, A.: Systematic literature review of software process capability/maturity models. In: Proceedings of International Conference on Software Process Improvement and Capability Determination (SPICE), Pisa, Italy (2010)

# Lessons Learned from Teaching Open Source Software Development

Becka Morgan[1] and Carlos Jensen[2]

[1] Western Oregon University, Monmouth, OR, USA
morganb@wou.edu
[2] Oregon State University, Corvallis, OR, USA
cjensen@eecs.oregonstate.edu

Abstract. Free/Open Source Software allows students to learn valuable real world skills and experiences, as well as a create a portfolio to show future employers. However, the learning curve to joining FOSS can be daunting, often leading newcomers to walk away frustrated. Universities therefore need to find ways to provide a structured introduction to students, helping them overcome the barriers to entry. This paper describes two courses taught at two universities, built around a Communities of Practice model, and the lessons learned from these. Suggestions and insights are shared for how to structure and evaluate such courses for maximum effect.

Keywords: Free/Open Source Software, Education, FOSS.

1 Introduction

Free/Open Source Software (FOSS) is an increasingly important part of the computing eco-system [1]. Teaching students how to participate in FOSS projects not only provides them with meaningful and highly marketable hands-on experience, it also potentially helps ensure FOSS communities have enough qualified developers to draw from to meet their needs. Supporting a growing and vital FOSS eco-system requires us to grow the pool of potential contributors. While working things out from first principles by yourself has been the traditional way of learning how to contribute to FOSS, this is a very inefficient model, and unlikely to meet growing needs.

Newcomers to FOSS often have a difficult time finding an entry point into a project. Barriers to entry include documentation that is incomplete and/or not up to date, no response from community members when questions are asked, and the need to learn a new set of tools in order to participate[2]. While braving the learning process and overcoming the many obstacles on your own is seen as a badge of courage, we believe there is ample room for improvement, and a need for a guided and more structured learning experience.

Beyond the economic/market arguments, as educators we have a duty to look towards the needs of our students, and how to best prepare them for the jobs of tomorrow. Experience with, and the ability to participate in FOSS development is becoming a critical skill, as companies increasingly adopt FOSS[3], and even when not using FOSS, look for FOSS experience in their hiring. According to David Heinemeyer Hansson, a partner at the software development firm 37signals, "Open source is a golden gift to the hiring process of technical people. It reduces the risk enormously by allowing you to sample candidates over a much longer period of time" [4].

As FOSS projects give students a unique way of developing real-world skills and experience, as well as providing prospective employers with tangible proof of their skills, it is important for universities to develop class-room models to teach more students how to participate in FOSS. Some universities have been experimenting with such courses, and this paper describes two different approaches to teaching students how to navigate the joining process and contribute to a FOSS project. One instructor focused on building on a collaborative "communities of practice" model, while the other on a more traditional classroom model. This paper compares and contrasts the two models different models, each used twice at different universities. It discusses the outcomes, lessons learned, and provides guidance to those contemplating developing their own courses on this topic.

2 Related Work

Work has been done looking at how to incorporate FOSS into CS curriculum, and even how to use FOSS projects as a way of recruiting students looking to have more real-world impact. One such effort is the Humanitarian FOSS project (HFOSS) [15]. This curriculum focuses on creating enthusiasm and engagement around participation in HFOSS, primarily among non-traditional CS students. Through this effort they seek to provide compelling use-cases and activities that will motivate a wide range of student groups, and be able to effectively leverage diverse backgrounds and experiences to solve real-world problems [9, 16].

The HFOSS project has also influenced other curriculum efforts, most notably the "Professors Open Source Summer Experience" (POSSE), sponsored by Redhat [5]. This effort is aimed at teaching college professors about FOSS, about FOSS tools, and how to introduce FOSS into their curriculum. This paper focuses on lessons learned from using the "go it alone" model, then incorporates information gained from attending POSSE to outline an improved curricular model designed to support students in overcoming the barriers to entry. In addition, there have been a number of efforts designed to produce curriculum around FOSS, such as the Teaching Open Source web community [6], which shares curriculum online. We build on both lessons and materials from POSSE, as well as our own personal experiences.

There is a preponderance of literature written about FOSS looking at who already contributes. Surveys have been conducted to look at who is participating in FOSS [7, 8, 9]. This provides us with a look at the demographics of contributors. There is also research into the motivation of FOSS developers [10, 11, 12, 13] looking at why developers volunteer time to work on FOSS projects. Additionally there is a body of work that considers the "newbie" experiences approaching and joining FOSS projects [14, 15, 16], a great deal of which focuses on the lack of diversity within FOSS developer populations and ways to affect change [15, 17, 18]. What is lacking from this body of work is a robust foundation for creating a pedagogical approach for curriculum that supports a platform for "newbies" to become involved.

To address this issue, both instructors decided to build on the Communities of Practice (CoP) theory, though to different degrees. Efforts aimed at supporting diversity in CS classrooms often builds on, implicitly or explicitly, the CoP framework (e.g. pairs programming [19, 20] and peer mentoring [21]). Research shows that women often turn away from CS, at least in part, because they are not shown what they can do with their knowledge in real life [22]. What they lack is the introduction to the CoP that they are joining.

Wenger defines communities of practice as "groups of people who share a concern or a passion for something they do and learn how to do it better as they interact regularly" [23]. This mirrors the FOSS way of doing things, and could serve as a theoretical framework for designing effective interventions. Wenger goes on to describe "three dimensions of the relation by which practice is the source of coherence of a community" [24]. These three dimensions are:

1. Mutual Engagement – practice exists in the relationships between people, developed as they engage in practice, whose meanings are negotiated with one another. Diversity is important in this engagement as each person brings different skills and competency to the practice.
2. Joint Enterprise – the enterprise the community is engaged in is defined by negotiation. The enterprise is thus defined by participants as it develops. This gives each participant a deep feeling of ownership of the enterprise and accountability to the community.
3. Shared Repertoire – as the enterprise is negotiated shared resources are developed. This includes artifacts, and routines, words, tools, stories, symbols, actions and concepts that are negotiated over time [24].

CS courses have traditionally focused on individual achievement and competition rather than cooperation, which does not reflect industry models, more focused on group accomplishments and team work [25]. The use of techniques such as pairs programming, a cooperative model of learning and development, has been proven beneficial [19]. Technical careers often rely on teamwork, and many companies use agile methods and extreme or pair programming.

Research conducted by POSSE participants has shown that it is both possible to teach students to participate in FOSS in a classroom setting, and that students find participating in Humanitarian FOSS (HFOSS) projects especially rewarding [26]. It is important now to create a body of literature that addresses the specific approaches that lead to success, failures and lessons learned, in order to provide a robust platform from which to develop more specific curriculum.

3   Learning Objectives and Pedagogic Approach

This paper details the experiences of two instructors teaching two independent courses in FOSS development at two separate universities over the course of two years.

The courses were taught using slightly different pedagogical styles, one providing more structure and guidance, as well as having a strong emphasis on in-class collaboration (Course A), and the second (Course B) following a more free-form individual-based structure. Both courses were built around the concept of building a CoP in the class, though to different degrees.

The goals of the two courses were the same: Providing students with the cultural and technical background to make a first contribution to a FOSS project, and the ability to identify future opportunities for contribution. The approach to meeting these goals was to provide a guided and structured process for navigating a joining process typically fraught with uncertainty and pitfalls. That said each instructor tailored the objectives to suit their target audience and class structure.

Course A was taught twice over two consecutive years at a small teaching university. Students in the first year course had junior and senior standing in Computer Science. A total of twenty-nine students were enrolled in the course. Students all worked on the same FOSS project and worked in teams to facilitate contribution. In the second year the course was opened up to upper class undergraduates in information systems and graduate students in management and information systems. There were a total of nine undergraduates and five students in the Masters program. Lessons were more guided and contributions were considered more broadly based on lessons learned from year one.

Course series B was taught twice over two consecutive years at a research university. The course was primarily offered to seniors and juniors in Computer Science, and followed a more traditional course model, where students picked their own projects to work on, and each student was responsible for their own coursework. There were twenty-four students in year one, and nineteen in year two.

Students were given some guidance in picking a project (more so in year 2 than year 1, when the lessons learned from the first year students were shared with the 2nd year students). This guidance was largely centered on specific projects to avoid or join, and organizational issues to look out for (active bug-tracker, up to date documentation, welcoming IRC/mailing list, etc.)

Both courses were designed to achieve the similar learning outcomes as follows:
1. Knowledge -
   (a) Give a definition of FOSS
   (b) Discuss the history of FOSS
   (c) Identify where to get answers to questions about Ubuntu. (A only)
2. Comprehension -
   (a) Explain the socio-political and technical workings of a FOSS project (B only)
   (b) Identify and summarize the workings of FOSS tools
3. Application
   (a) Map a generic path for joining an FOSS
   (b) Make use of FOSS tools (e.g., version control, bug tracker, IRC)
4. Analysis
   (a) Identify a project to contribute to   (B only)
   (b) Identify potential resources and key stakeholders (B only)
   (c) Identify areas of participation using current skills (A only)

(d) Separate research of the areas discovered to be addressed by individual group members (A only)
5. Synthesis
   (a) Assemble individual research into a plan to gain entry into an aspect of Ubuntu (A only)
   (b) Devise a plan as a group to contribute, using additional input from Ubuntu mentors (A only)
   (c) Make three contributions to a project (B only)
6. Evaluation
   (a) Explain to other groups how to complete tasks in one area of the Ubuntu project (A only)
   (b) Document contribution, and present to class (B only)

4 Course Design

4.1 Course A

Course A was designed around cooperative learning. The Ubuntu project was chosen for all students to work on to remove uncertainty and facilitate in-class collaboration. Ubuntu is a community that is known to be welcoming, has extensive documentation, and the availability of mentors. In addition to mentors, students also had access to weekly meetings of Ubuntu contributors and a prerelease Global Jam toward the end of the term. The Global Jam is an Ubuntu coordinated event uniting community members worldwide to improve the Ubuntu project. Students had several choices for how to contribute, including documentation, design, development, bug triage, and testing. Students were assigned to groups based on their area of interest. Each student had a mentor assigned to them.

As a co-operative classroom the course was taught in a lab, giving students ample hands-on experience. There were no lectures in this course; rather time was spent in discussion. If a group hit a barrier they could not overcome, they either sought help from other groups, mentors or the instructor. The term was divided with the beginning focused on history and culture, both of FOSS in general and Ubuntu specifically, followed by an introduction of tools used in FOSS, including IRC, version control, and bug-trackers. This work was used as the foundation for contribution.

After the first four-week introduction, students began to focus on contributions. As a starting point all students were assigned to bug triage, documentation, or testing. Students were required to make at least one contribution to the Ubuntu project. Groups worked to gather information from the Ubuntu documentation and their mentors to complete this assignment. Students were assessed based on participation in their team (30%), attendance (30%), and contributions to Ubuntu (40%). The final piece took into consideration the difficulty of the contribution. Fixing a bug title did not carry as much weight as testing code or creating quick lists (a dropdown menu in Ubuntu Unity giving quick access to common tasks for applications).

4.2 Course B

Course B was designed to more closely mirror the typical lecture-based classroom experience, as well as give students more flexibility in what they wanted to get involved in. Self-motivation, the desire to scratch your own itch, is an important part of what drives FOSS contributors. This freedom of course comes at a cost; there are no guarantees that students won't pick hostile or non-productive projects, and the guidance and support that they can receive from the rest of the class and instructor are limited because each project is likely to differ in terms of tools used, customs, and availability of documentation. We chose to use this structure for this class because the student population was deemed to be more mature and experienced, and thus potentially better able to deal with potential setbacks.

Despite the divergence in projects, and that each student worked independently, there was an important social component, building on the CoP theory. There were weekly oral status reports before the whole class, and more lengthy bi-weekly experience reports and discussions. Students shared their progress, lessons learned, and pitfalls encountered, and gave each other advice on how to proceed. This helped build shared repertoire, and mutual engagement, key components to a CoP.

Lectures were based on materials from *The Cathedral and the Bazaar* [27] and *Producing Open Source Software* [28]. Most of the lecture time was aimed at providing students with the required background, including an understanding of process, and the tools used in FOSS. A lot of the technical material was front-loaded to get students up and running quickly, and the second half of the course was designed around discussions of open problems in FOSS and speakers talking about their projects.

Students were evaluated through a midterm and final exam focusing on the historical, political and licensing frameworks of FOSS (30% total grade). They had small assignments distributed through the early part of the term aimed at helping them get familiar with the organization of a particular project, and the tools used in said project (org overview and mapping of resources; use of IRC and mailing lists; and code repository and bug tracking). This accounted for 20% of their grade.

The remaining 50% was distributed over 3 contributions of the students choosing to their project of choice (bug report, documentation, testing or code patch, no more than 2 of any one). Contributions had to be accompanied by a report reflecting the importance of the work, and the process followed to create and submit it to their project. Students had bi-weekly project reports and updates to their classmates, which kept everyone up-to-date, and helped disseminate best practices and pitfalls.

Because of the short duration of a term and the long review process associated with many projects, contributions were graded by the amount of effort and whether the process was adequate rather than whether the contribution was ultimately accepted. Acceptance was nice, but not always achievable to a true novice on the first try.

5 Results

The final analysis of course A showed that in year one 25 out of 30 students (83%) made some contribution to the project. Out of those 25 students 16 (53% of the total) made more than five contributions of varying degrees of difficulty. Additionally 3 out of 5 women and 4 of 6 non-white were part of the groups that had the highest performance levels. Analysis of year two shows that 11 out of 14 students (79%) made contributions to the project and of those 11 students 5 (45%, of the total) made more than five contributions. Given the difference in skill sets due to the mix of CS, IS and MIS, the level of contribution was not considered in year two.

Course B similarly had a high degree of success, with 19 of the 24 students completing their three contributions in a satisfactory manner in year one (79.2% of students), and 17 of 19 students (89.5% of students) in the second year. Of those who did not complete all three assignments, all completed at least one contribution in year one, and two in year two. For the second year, student evaluations indicated that this was one of the most worthwhile courses the students had completed, and more than half the students indicated that they intended to continue working on the project they had chosen. For the first year, evaluations were mixed, primarily due to frustration with the projects they had selected and issues of non-responsiveness and lacking or misleading documentation, and fewer than a quarter of students expressed an interest in continuing their work on the projects after the end of the class.

6 Discussion and Lessons Learned

While it is difficult to objectively evaluate the effectiveness of the courses we designed, other than in terms of meeting their learning objectives, we judge them to have been a success. Joining a FOSS project for the first time is a very time-consuming and uncertain process, and for most students, either course helped them successfully navigate the learning process in a very limited amount of time.

As should be evident from the description of the courses presented, a significant amount of the learning was still self-directed; the students had to figure out what to do for themselves. What the course did provide was a social and technical context for doing so. The class gave them a goal, helped explain some of the organizational and cultural issues they were previously unfamiliar with, and the understanding that this process is difficult, and that it is OK to be lost. With this context, the students were largely able to navigate the process with some minimal support and encouragement.

Though we only have anecdotal evidence to support this, we found that the social context, the CoP framework that we sought to build, was immensely valuable. While many students initially were reluctant to admit problems or failures, and indeed would likely have struggled in silence with these until they succeeded or gave up, by the midpoint of the course there was a definitive feeling of camaraderie among the students. They shared their frustrations with each other openly, and were able to offer advice and coaching to each other unprompted. Forming in-class teams or not does not seem to have been a factor for success or failure.

That said, we did experience some significant problems in our courses. The openness of course A (make any number of contributions of any kind) coupled with the size of the Ubuntu project turned out to be a significant obstacle. As a result students reported being overwhelmed and unable to find a place to begin. It is worth noting that students who interacted with their mentors and each other on a regular basis found it easier to contribute to the project. It became apparent that the course needed more structure and direction. Students also needed to be taught how to use their mentors to gather information. Just providing mentorship did not ensure that students knew what to ask or how to get started talking to their mentors.

Course B was in some ways more open than Course A, but students were directed to either smaller projects, projects they already had experience with, or projects that had an established mentor network. The expectations in terms of contributions were better defined however. Fewer students reported being overwhelmed as a result.

Because of the openness of course A, evaluation was difficult. Grades were based on participation, contribution, and attendance. The difficulty evaluating students stemmed from not having clear guidelines and expectations. This, coupled with the lack of structure and overwhelming size of the project, pointed to a solution that would address all three problems. Using a smaller project written in a language students had experience with would provide students with a more moderate learning curve. The course needed assignments to provide the means to gain access to the project, but also as a means to evaluate students' work. Course B on the other hand turned out to be much easier to evaluate, and the exams and minor assignments helped students feel less worried about their grades.

The end result of using Ubuntu was the realization that the project was too large. Although there were many places to participate, and the community was very welcoming, most of the students reported being overwhelmed with the documentation of the project. While the documentation was exhaustive, the sheer volume made it difficult for students to find answers to their questions and a place to start. Fifteen out of twenty-seven students listed being overwhelmed by the amount of documentation and/or finding an entry point into the project because it was so large.

Not picking a project for students however led to more work for the instructor, and some additional uncertainty among students early in the term, but in the second year we developed guidelines for selecting better projects, and students were more devoted to their tasks.

These courses will continue at both universities, and the curriculum is available for others seeking to adopt or design their own curriculum <links to be shared after review>.

7   Conclusions

Using FOSS in higher education serves both the students – by providing real world experience, as well as the FOSS community – by growing a pool of potential developers. Courses that emphasize hands-on experience and the completion of real contributions to real FOSS projects were able to achieve a very high success rate.

Following a CoP model, whether strictly or loosely interpreted, contributed to this success, and helped students overcome the confusion and frustration of dealing with an often unstructured learning challenge.

References

1. Deshpande, A., Riehle, D.: The total growth of open source. Open Source Dev. Communities Qual., 197–209 (2008)
2. Shibuya, B., Tamai, T.: Understanding the process of participating in open source communities. In: ICSE Workshop on Emerging Trends in Free/Libre/Open Source Software Research and Development, FLOSS 2009, pp. 1–6 (2009)
3. Trapasso, E., Vujanic, A.: Accenture Newsroom: Investment in Open Source Software Set to Rise, Accenture Survey Finds (2010)
4. Hansson, D.H.: Reduce the risk, hire from open source (2005),
   http://www.loudthinking.com/arc/000505.html
5. Ellis, H.J., Chua, M., Hislop, G.W., Purcell, M., Dziallas, S.: Towards a Model of Faculty Development for FOSS in Education
6. Teaching Open Source,
   http://teachingopensource.org/index.php/Main_Page
7. Ghosh, R.A., Glott, R., Krieger, B., Robles, G.: Free/libre and open source software: Survey and study. Maastricht Economic Research Institute on Innovation and Technology, University of Maastricht, The Netherlands (June 2002)
8. David, P.A., Waterman, A., Arora, S.: FLOSS-US: the free/libre/open source software survey for 2003. Stanf. Inst. Econ. Policy Res. (2003),
   http://www.Stanf.Edu/group/floss-Us (accessed September 20, 2004)
9. Lakhani, K., Wolf, B., Bates, J., DiBona, C.: The boston consulting group hacker survey. Boston Consult. Group (2002)
10. Hars, A., Ou, S.: Working for free? Motivations of participating in open source projects. In: Proceedings of the 34th Annual Hawaii International Conference on System Sciences, p. 9 (2001)
11. Hertel, G., Niedner, S., Herrmann, S.: Motivation of software developers in Open Source projects: an Internet-based survey of contributors to the Linux kernel. Res. Policy 32, 1159–1177 (2003)
12. Roberts, J.A., Hann, I., Slaughter, S.A.: Understanding the motivations, participation, and performance of open source software developers: A longitudinal study of the apache projects. Manag. Sci. 52, 984 (2006)
13. Ye, Y., Kishida, K.: Toward an understanding of the motivation of open source software developers. In: Proceedings of the 25th International Conference on Software Engineering, pp. 419–429 (2003)
14. King, S., Kuechler, V., Jensen, C.: Joining Free/Open Source Software Communities An Analysis of Newbies' First Interactions on Project Mailing Lists. Presented at the 2011 44th Hawaii International Conference on System Sciences, January 1 (2011)
15. Kuechler, V., Gilbertson, C., Jensen, C.: Gender Differences in Early Free and Open Source Software Joining Process. In: Hammouda, I., Lundell, B., Mikkonen, T., Scacchi, W. (eds.) OSS 2012. IFIP AICT, vol. 378, pp. 78–93. Springer, Heidelberg (2012)
16. Park, Y.: Supporting the learning process of open source novices: an evaluation of code and project history visualization tools (2008)

17. Byfield, B.: Sexism: Open Source Software's Dirty Little Secret - Datamation, http://www.datamation.com/osrc/article.php/3838186/Sexism-Open-Source-Softwares-Dirty-Little-Secret.htm
18. Levesque, M., Wilson, G.: Women in software: Open source, cold shoulder. Softw. Dev. (February 20, 2005), URL Consult., http://www.Sdmagazine.Com/documents/s=9411 (2004)
19. McDowell, C., Werner, L., Bullock, H.E., Fernald, J.: Pair programming improves student retention, confidence, and program quality. Commun. ACM 49, 90–95 (2006)
20. Nagappan, N., Williams, L., Ferzli, M., Wiebe, E., Yang, K., Miller, C., Balik, S.: Improving the CS1 experience with pair programming. ACM SIGCSE Bulletin, 359–362 (2003)
21. Cohoon, J.M.G., Gonsoulin, M., Layman, J.: Mentoring computer science undergraduates. Hum. Perspect. Internet Soc. Cult. Psychol. Gend. 4, 199–208 (2004)
22. Margolis, J., Fisher, A.: Unlocking the clubhouse. MIT Press (2002)
23. Wenger, E.: Communities of practice: A brief introduction (2006) (retrieved October 1, 2008)
24. Wenger, E.: Communities of practice: Learning, meaning, and identity. Cambridge Univ. Pr. (1998)
25. Howell, K.: The experience of women in undergraduate computer science: what does the research say? ACM SIGCSE Bull. 25, 1–8 (1993)
26. Morelli, R., Tucker, A., Danner, N., De Lanerolle, T.R., Ellis, H.J., Izmirli, O., Krizanc, D., Parker, G.: Revitalizing computing education through free and open source software for humanity. Commun. ACM 52, 67–75 (2009)
27. Raymond, E.S.: The cathedral and the bazaar: musings on Linux and open source by an accidental revolutionary. O'Reilly & Associates, Inc. (2001)
28. Producing Open Source Software, http://producingoss.com/

A Successful OSS Adaptation and Integration in an e-Learning Platform: TEC Digital

Mario Chacon-Rivas[1] and Cesar Garita[2]

[1] TEC-Digital,
Costa Rica Institute of Technology (TEC), Cartago, Costa Rica
machacon@itcr.ac.cr
[2] Computing Research Center, School of Computer Science,
Costa Rica Institute of Technology (TEC), Cartago, Costa Rica
cesar@itcr.ac.cr

Abstract. E-learning projects in many universities are focused on adapting or installing a software platform to upload teaching materials and sometimes to open discussion forums. However, it is totally possible to extend the learning management system (LMS) as a complete service platform for students and instructors including more advanced services. This paper shows the progressive integration of services and applications in TEC Digital as the open source e-learning platform of the Costa Rica Institute of Technology. This integration experience could be used as a case of study for other universities.

1 Introduction

In 2008, the Costa Rica Institute of Technology (TEC) started a project called TEC Digital to renew its LMS (Learning Management System) platform. The general objective of TEC Digital was to incorporate ICT in the development of teaching activities at TEC [1]. One of the main requirements then was that it should be open source [2]. Thus, the open source strategy that guides the architecture of TEC Digital, offers a high degree of extensibility which has led to the development of novel tools in the areas of instructional design, m-learning, adaptive learning, business intelligence, usability, and competence management, among others.

This paper shows the progressive integration of services and applications in the TEC Digital e-learning platform. This work could be used as case of study for other universities facing the problem of integrating e-learning technologies using OSS.

2 Related Work

During the last two decades, many papers can be found concerning the comparison and adoption of LMS in universities [3][4]. There are different justifications for adopting one LMS or the other, based on given technical criteria [5]. In particular, there are several studies about Free/Libre Open Source Software (FLOSS) adoption in public sector and in e-learning. For instance Rossi et al, in [6], enumerated several *individual*, *technological*, *organizational* and *environmental* factors to be considered in OSS adoption. Also, in [7] the FLOSS community presents a useful study of adoption models and myths around open source solutions. In [8], a comparative analysis is presented regarding FLOSS LMS, and one of the main conclusions is that despite technical functionalities, the final decision is greatly affected by end user needs.

In terms of open source LMS architecture approaches, there are several works focusing on extensibility and service integration. In [9], an extendable open source architecture of e-learning systems is presented, based on core components and optional extensions. In [10], a service-oriented architecture for LMS is discussed to support interoperability between LMS and different systems and databases. The architecture of TEC Digital follows these principles of extensibility and interoperability in order to support the development of novel components, as described in the next section.

3 Integrating Services in TEC Digital

The process of adaptation and integration of the full e-learning system at TEC has involved the following main tasks or stages:
   (1) Adoption of .LRN LMS within the university.
   (2) Integration of internal information sources into .LRN.
   (3) Integration of complementary services and applications.
   (4) Development of novel complementary components.

These tasks are briefly explained in the following subsections.

3.1 Adopting .LRN

In TEC, the adoption of the LMS was based on a comparison study between different platforms such as Moodle, Sakai and .LRN. In the end, .LRN was selected mostly due to its virtual community management, portal creation and interface design. The main issues compared against those LMS included: technical staff, interface design, adaptability to our internal organization, community use and development support. Once the LMS was chosen, a pilot plan was started involving a group of instructors from a few schools. Then, in 2009 .LRN was officially adopted as institutional platform and it has been consolidated through daily use.

3.2 Integration of Internal Information Sources

It is very common that universities have the LMS installed on one platform and the administrative and academic information located on other different platforms. In TEC case, the academic and administrative information systems are based on a Microsoft platform. The .LRN architecture is based on open source solutions including PostgreSQL, OpenACS, AOLServer. In order to integrate information services from the internal university sources (e.g. student admission and registration department), a service-oriented architecture was implemented. This integration architecture allowed the addition of several features and tools such as: automatic users account creation from registry database, student/instructor profiles, and communities matching the internal organization.

3.3 Integration of Services and Applications

The TEC academic community requires several services and applications offered by third parties to support teaching and regular activities. Some of those services are: Web2Project, Munin, Limesurvey, CmapServer, OJS, GIT, Dspace. The integration of these components has been mostly done using OpenLDAP and integrating some functions internally into the LMS.

3.4 Development of Novel Components

Besides integrating existing services and components, TEC Digital has the objective of complementing (improving) the LMS platform with some advanced functionalities or tools for students and teachers as well as university managers and collaborators. Some of the components developed "in-house" that have been successfully integrated with .LRN include: Instructional Design Generator (course planning), Mobile Course (app for course information access), and Learning Activities Manager (evaluation). Please notice that the adoption and integration of the services and components described in this section, demands a highly extensible system architecture, based on well-defined layers, web services and protocols.

Through the described services and components, TEC digital manages around 12125 students, 1026 instructors, 148 virtual communities, 4481 courses, and 8500 user accesses by day.

4 Conclusions

Some of the major benefits for our university generated by TEC Digital open source solutions include: reduced software license costs, seamless integration of university systems, capacity building in open source solutions and perhaps, the biggest benefit of OSS investment is that the organization is the "owner" of its products and solutions.

On the other hand, there are some considerations regarding the use of OSS that should not be underestimated: the use of OSS in an organization and in particular at universities, must be carefully planned in several dimensions; the technical skills required for technicians supporting OSS platform can be different than in other platforms; the use of OSS does not mean no investment, the organization must invest in support, adaptation, integration and others, and finally, the definition and coordination of the technical team can be quite complex.

References

[1] Garita, J.A., Alpízar, I., Chacón-Rivas, M.: OpenACS/dotLRN integration with ITCR platform. In: 8th OpenACS/dotLRN Conference, Cartago, Costa Rica, pp. 27–35 (2009)
[2] Garita, C., Chacón-Rivas, M.: TEC Digital: A case study of an e-learning environment for higher education in Costa Rica. In: Proceedings of ITHET 2012. IEEE Proceedings ITHET, pp. 1–6 (2012)
[3] Ozkan, S., Koseler, R.: Multi-Dimensional Evaluation of E-Learning Systems in the Higher Education Context: An Empirical Investigation of a Computer Literacy Course. Presented at the 39th ASEE/IEEE Frontiers in Education Conference, pp. 1–6
[4] Aberdour, M.: Open Source Learning Management Systems. In: EPIC (2007)
[5] Moreno, P., Cerverón, V.: Plataforma tecnológica para potenciar los procesos de enseñanzaaprendizaje: desarrollo en la Universitat de València basado en software libre y colaborativo. In: Proc: SIIE 2006. VIII Simposio Internacional de Informática Aplicada a la Enseñanza (2006)
[6] Rossi, B., Russo, B., Succi, G.: Free/Libre Open Source Adoption in the Public Sector: Current State and Lessons Learnt
[7] FLOSSMetrics project, http://flossmetrics.org/ (accessed October 13, 2010)
[8] Fernandes, S., Cerone, A., Barbosa, L., Papadopoulos, P.: FLOSS in Technology-Enhanced Learning, http://mlab.csd.auth.gr/index.php/gr/publications?view=publication&task=show&id=44 (accessed January 24, 2014)
[9] Khan, M.A., UrRehman, F.: An Extendable Open Source Architecture of e-Learning System
[10] Jabr, M.A., Omari, H.K.: E-learning management system using service oriented architecture. J. Comput. Sci. 6(3), 285 (2010)

Smart TV with Free Technologies in Support of Teaching-Learning Process

Eugenio Rosales Rosa, Abel Alfonso Fírvida Donéstevez,
Marielis González Muño, and Allan Pierra Fuentes

University of Informatics Sciences, School 1, Free Software Center
San Antonio de los Baños Highway, Km 2 ½, Torrens, Boyeros, Havana, Cuba
{erosales,aafirvida,mmuno,apierra}@uci.cu,
http://www.uci.cu

Abstract. The digital divide created between Cuba and the rest of the world has forced us to use alternative technologies in order to preserve and strengthen the achievements of the Revolution in the field of education. One of the actions undertaken in this regard consists of making audiovisual equipment and media become a supplementary element of the teacher's educational work, and thus ensuring the rational use of the aforesaid media. This paper shows how to use a new trend of information technology and communications, using hybrid or smart TVs. This low-cost solution for low energy consumption, conceived as part of the educational process at all levels of the Island, provides some technical aspects and also shows, in the outline, some other ideas for incorporating this technology into the teaching-learning process. The results of laboratory tests are likewise shown.

1 Introduction

The birth of Information and Communication Technologies *(ICT)* has brought about deep changes in the structure of teaching aids, by adding some new tools and changing the existing, more traditional methods and techniques. These changes have also influenced the way of using teaching aids, since they have contributed to the optimization of newer techniques involved in the process of learning, granting more access to it *(Bravo, 2004)*.

Television is one of those teaching aids which have since long transformed education inside and outside the school. By the early Seventies, cartoons made for children in the Soviet Union played a major educational part, and the year 1985 would see the emergence of Discovery Channel in the United States, thus becoming the first worldwide learning channel.

In Cuba television sets are used as teaching aids since 1961, when they were first use during the literacy campaign was on its way. Later on, by the year 1968, a series of televised programs was conceived with the aim of supporting classes in elementary and high schools.

During the 2000 – 2001 school year, the initiative of using television as a teaching aid was taken up again and in some elementary, junior and senior high schools, a television set was used for every 100 students. At that time, the materials for teaching students only reached the audience as part of other programs broadcast by the national Cubavisión channel. All along the 2001 - 2002 school year, each of the 9970 schools *(Cuba, 2010)* received a TV set per classroom. The total figure of televisions reached the number of 109.117 and they came along with some other 40.858 video cassette players *(VICENT, 2005)*. This major boost for education was firmly intended to extend the range of teachers' options to enrich their classes, by replacing expensive teaching aids with audio-visual materials in a more interactive way. Two more learning channels were conceived, with the purpose of broadcasting televised classes and educational multimedia that would back up the teaching process.

The University of Computer Sciences (UCI) was inaugurated in the year 2002. Within the university, as part of the back-up equipment and of the upgrading of the educational system, at least two TV sets and a computer were placed in each classroom, as well as other devices allowing the connection with the aforementioned machines. Different television channels were opened for the broadcast of televised classes, consultation and other educational materials. Besides, there was a website that granted access to any of those resources at anytime from a computer.

From what has been said, one can conclude that since the economic situation of the country is not at its best, it would be advisable to look for cheaper alternatives (so as to be able to replace and/or repair the computers in case of damage) and a lower energy consumption (to grant a rational use) for the technologies used in classrooms all over the country.

And then? How could a teaching aid based on ICT be used as a supplement to the teaching process and be likewise as cheap and energy-saving?

The aim of the present research is to design a teaching aid based on ICT, using low-cost and energy-saving elements that can be used as a supplement to the teaching process.

2 Smart Television

A smart television (STV) is made up by a television designed to process and store data like a normal computer, allowing users to watch their favorite programs by request, without being overwhelmed by the flood of repetitive TV commercials. Instead, commercials appear in the guise of inserted messages or *gadgets* or over the broadcast itself. This kind of equipment may also take in games, applications and be able to interact with Internet through the e-mail, instant messages or Web surfing *(Kovach, 2010)*.

2.1 Background

By the year 2010, the enterprises like Google, Intel and Sony got together to announce to the world the news about a revolutionary appliance, the smart television, a technological novelty based on digital television. Yet, the social object for which STV were designed is much distant from the conception of a didactic teaching aid. Its introduction in Cuban schools may be favorable for the teaching-learning process because of the following reasons:

- The STV is a technological evolution of video cassette players, computers and televisions presently used at all levels of instruction in the country.
- The educational software would be easily inserted in the learning process as didactic materials.
- The STV would facilitate the use of the network and grant access to Internet in order to get information and the didactic resources.
- It could be put into service using low-cost and energy saving components.

The new potential of smart televisions is due to the inclusion, in their circuits, of small computers like the ones used in mobile phones, tablets or even in portable computers. There are several transnational companies which are presently developing this kind of TV; among them are the world-known LG, Google and Samsung. There are also important projects for Free and Open Code Software (FOSS) which have switched from desk computers to television sets, such as GNU/Linux Ubuntu and even some others which had initially foreseen a foray into this kind of media like the Android project.

On the other hand, there are some hardware projects whose aim is to have small computers plugged in to the television. These would be extremely cheap since they don't even come with a chassis. Most likely, there isn't any other solution as popular as Raspberry Pi. Its manufacturer describes it as: (…) a computer whose size is that of a credit card that is plugged in to a TV and a keyboard. It's a PC that can do many a thing like a desk PC: spreadsheets, text processing and games. It can also play high-definition videos *(Raspberry Pi, 2011)*.

After Raspberry Pi started marketing computers the size of a card, several other brands have arisen, having a higher quality with regards to the processing potential and other technical features. Among the most promising ones, being in the core of the present research, are Odroid and APC 8750, these two have an outstanding feature: they are sold with the operating Android system already installed. In response to the primary goal set by this research, the use of the Raspberry Pi board was decided, mainly because it is cheaper than the two other options and presents an output device for the RCA video which is at the same time the standard input of television since the late Fifties.

A chart showing a comparison of the abovementioned boards is found below:

Table 1. Comparison of boards

Board	Price	CPU	RAM	Video
R. Pi (B)	$35	700 MHz	512MB	HDMI/RCA
Odroid (U2)	$89	1.7GHz	2GB	HDMI
APC (8750)	$49	800 MHz	512 MB	HDMI/VGA

2.2 Operating System

Since the year 2008, different free and/or open code projects have intended to provide an operating system for both types of devices. By the late 2009, Google released the first version of Android's operating system, which was backed up by the giant Internet transnational company and by Open Handset Alliance (OHA)[1]. Since 2010, the home page of Android's developers has announced that they are working on the transfer of this platform on to televisions.

Curiously, for Android, it was decided to have the architecture of a conventional GNU/Linux distribution changed. Why? Any GNU/Linux distribution has in its default settings some applications developed by using programming languages such as Perl, Python, Bash y C/C++, and also developed by using different graphic libraries like Ncurses, Gtk+ and/or Qt. This diversity causes a certain inefficiency that is not perceived or is just in any case of minor importance in a desk computer or telephone with limited resources. Standardizing development technologies (Java y C/C++) brought about the birth of powerful tools like SDKs and IDEs integrations like Eclipse, an incomparable phenomenon in the world of FOSS.

These tools have also given rise to some applications for Android in the last four years, surpassing the number of applications of any GNU / Linux distribution. Therefore, for the present research we propose the use of Android's operating system, for its maturity, and for it has a large number of applications and a version for Raspberry Pi.

2.3 Known Results

The STV has more interactive elements than presentations or slideshow based on static images do, since it can even contain live videos. For classes and teachers working in the field of information and communication, the television is a good chance to use educational applications.

It is advisable that the physical duration of the object ranges from 10 to 20 minutes. On its own side, the length of the student's learning period does not have a definite pattern, because it depends on her/his own skills *(FONDEF-APROA, 2005)*.

Applications for the STV are compared, in this case, to the objects of learning, since the structure of an application for Android (the operating system chosen) ensures all the good practices that have been previously described. Besides, these applications make the use of templates easier, as well as the television's design, since they save time and resources while generating objects and putting them into sequence in a similar learning context. The use of templates is favorable for the design of the object and also for the process of understanding of the contents by the students, who will rely on a standard-format object *(FONDEF-APROA, 2005)*.

---

[1] Association of 84 companies whose aim is to develop a platform that allows users to interact with mobile systems in a better and cheaper way (Open Handset Alliance), among its members are to be found: LG, Aser, ASUS, ALCATEL, ZTE, Huawey, DELL, Fujitsu, HTC, Samsung, inter alia.

Below, a chart shows a comparison of the energy consumed by some teaching aids:

Table 2. Energy comparison among teaching aids

Audiovisual aids	Components' Power (W)				
	Television	Computer	Monitor	Video/DVD	Total
Use of video/DVD	75	-	-	20	95
Use of computer	75	300	60	-	435
Smart television	75	4	-	-	79

3 Conclusions

At the completion of the present research, a smart television set was designed, being able to work as a teaching aid and having the following features:

- Low production cost, since it can be manufactured by placing a 35 USD computer in the interior of the TV sets which are already in use in classrooms all over the country.
- Low energy consumption, since this brand-new device would only consume 4W more than a regular TV set.
- Lined up with the latest developments of information and telecommunication technologies, it would use Android as the fastest-expanding and most profitable operating system at present.
- Also lined up with the most recent information and telecommunication trends in education, it would propose the use of applications as subjects of study.

As from now on, this team will focus on introducing some samples of the teaching aid in real classrooms, in order to validate its proposal and to put this technology in service all over the nation as soon as possible.

Furthermore, the thread of software and hardware components used by the aforesaid system will be followed closely, since production costs tend to be lower all the time and it is likely that newer technologies will arise, making STV cheaper and more rational for Cuban education.

References

Ávila, M.A.A.: Diez preguntas sobre el ahorro de energía eléctrica (2010), http://www.cubasolar.cu/biblioteca/energia/Energia33/HTML/articulo03.htm (retrieved March 30, 2013)

Bravo, J.L.R.: Los Medios De Enseñanza: Clasificación, Selección y Aplicación. Revista Pixel-Bit 24 (2004), http://www.sav.us.es/pixelbit/pixelbit/articulos/n24/n24art/art2409.htm (retrieved)

Oficina Nacional de Estadisticas de Cuba, Anuario Estadístico de Cuba 2010 (2010), http://www.one.cu/aec2010.htm (retrieved March 29, 2013)

Fondef-Aproa, Manual De Buenas Practicas Para El Desarrollo (2005),
http://www.aproa.cl/1116/articles-68370_recurso_1.pdf (retrieved)

Kovach, S.: What Is A Smart TV? - Business Insider (December 8, 2010),
http://www.businessinsider.com/what-is-a-smart-tv-2010-12
(retrieved March 29, 2013)

Linux Infra Red Control. (n.d.). How to connect an IR receiver to an audio card,
http://www.lirc.org/ir-audio.html (retrieved March 29, 2013)

ODROID | Hardkernel (2012),
http://www.hardkernel.com/renewal_2011/products/prdt_info.php
(retrieved March 29, 2013)

Raspberry Pi | An ARM GNU/Linux box for $25. Take a byte! (2011),
http://www.raspberrypi.org/ (retrieved March 29, 2013)

Vicent, M.: Revolución en las aulas de Cuba | Edición impresa | EL PAÍS (Abril 2005),
http://elpais.com/diario/2005/04/04/educacion/
1112565605_850215.html (retrieved)

Barriers Faced by Newcomers to Open Source Projects: A Systematic Review

Igor Steinmacher[1], Marco Aurélio Graciotto Silva[1], and Marco Aurélio Gerosa[2]

[1] DACOM, UTFPR Campo Mourao, PR, Brazil
{igorfs,magsilva}@utfpr.edu.br
[2] IME, USP Sao Paulo, SP, Brazil
gerosa@ime.usp.br

Abstract. To remain sustainable, some open source projects need a constant influx of new volunteers, or newcomers. However, the newcomers face different kinds of problems when onboarding to a project. In this paper we present the results of a systematic literature review aiming at identifying the barriers that a newcomer can face when contributing to an Open Source Software project. We identified and analyzed 21 studies that evidence this kind of problem. As a result we provide a hierarchical model that relies on five categories of barriers: finding a way to start, social interactions, code issues, documentation problems and newcomers' knowledge. The most evidenced barriers are newcomers' previous technical skills, receiving response from community, centrality of social contacts, and finding the appropriate way to start contributing. This classification provides a baseline for further researches related to newcomers onboarding.

1 Introduction

Some open source software (OSS) communities composed of volunteers from different parts of the globe contributing and collaborating. According to Qureshi and Fang [14], motivate, engage, and retain new developers is the way to promote a sustainable amount of developers in a project. However, newcomers often face difficulties and obstacles when onboarding to a project [8]. This obstacles can lead newcomers to give up their collaboration. Therefore, a major challenge for OSS projects is to provide ways to support the joining of newcomers.

To reduce these problems, newcomers generally post questions and request help to choose their tasks in forums and mailing list or send emails to developers who have central roles in the project (e.g. owners, project leaders) [13,22]. However, receiving replies that do not offer guidance or unpolished answers can result in newcomers to give up contributing [18]. Given this scenario, it is important to understand the OSS newcomers needs. This understanding may enable the creation of mechanisms and tools to offer a better support for newcomers.

The objective of this research is to identify the barriers faced by newcomers when onboarding to OSS projects. Onboarding is the stage in which an outsider decides to contribute to a project. Onboarding is highly impacted by a steep learning curve as well as reception and expectation breakdowns [17].

In this paper, the methodology chosen to collect these issues is the Systematic Literature Review. From the best of our knowledge, there is no study that directly focused on problems or barriers encountered by newcomers of Open Source Software projects. On the other hand, several articles report these barriers as a side product of the studies. Thus, knowledge is spread across the literature. This study main contribution is aggregating the barriers evidenced by different studies and creating a model with them.

2 Research Method

To perform our systematic review, we defined the following question: What are the barriers that influence newcomers' onboarding to OSS projects? By answering this question, we aim to capture the barriers that a newcomer can face when contributing to an OSS project. We are not interested in newcomers' motivation to join a project, but in the issues they can face after deciding to contribute.

After using different synonyms and combinations to refine our search, the query presented below was used to retrieve the studies from the following digital libraries: ACM, IEEE, Scopus and Springer Link. These libraries were selected because they index the most relevant venues of computer sciences, mostly written on English, they support searching using boolean expression and provide access to the complete text of the paper. We also consulted specialists for conferences, workshops, journals, and websites that could provide relevant studies for our research. However, no new source was added after their advices.

```
((OSS OR "Open Source" OR "Free Software" OR FLOSS OR FOSS) AND
(newcomer OR "joining process" OR newbie OR "new developer" OR "new contributor" OR "new
member" OR "new committer" OR novice OR beginner OR "potential participant" OR retention
OR joiner OR onboarding))
```

For each selected paper obtained from digital libraries, we conducted snowball sampling checking if the authors of the selected studies published other relevant studies not retrieved from the Digital Libraries. We checked their profiles in ACM, IEEE, DBLP, and personal homepages (when available).

We considered for selection the papers that were available for download, written in English, that dealt with newcomers onboarding in open source software projects, that presented experimental results, and that were published in journals or workshop/conference proceedings.

Subsequent to the definition of the primary studies list, the researchers read the full documents. To classify the barriers we followed an "inductive coding" approach [21], which is widely applied in qualitative studies of different knowledge areas. In this kind of approach, the evaluator identifies text segments that contain meaningful units and creates a label for a new category to which the text segment is assigned. Afterwards, connections between the codes are identified and they are grouped according to their properties to represent categories.

The results of the selection and screening are as follows. After running the query on the digital libraries systems, we got 291 candidate papers. For each paper, title, abstract and keywords were analyzed by two independent researchers.

In a consensus meeting, we came to 33 candidate papers. We checked other papers published by the authors of these 33 candidate studies, finding 20 other candidate papers. After analyzing the abstract of these papers we selected 9 relevant papers, coming to a total of 42 candidate papers. After further analysis, a total of 21 papers were considered relevant to this review and were considered to extract relevant data.

3 Barriers Faced by Newcomers

The main purpose of this systematic review was to find what are the barriers faced by newcomers to open source projects reported by the literature. For each selected study, we analyzed any barrier reported that was empirically identified or evaluated. We extracted the barriers from the selected studies, and organized them as a hierarchy of barriers, as shown in Figure 1. The figure presents five categories: Social Interactions, Finding a Way to Start, Documentation Problems, Code Issues, and Newcomers' Knowledge.

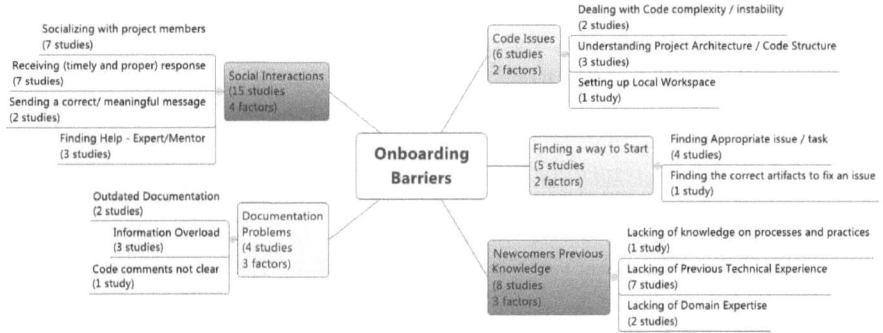

Fig. 1. Hierarchical map of barriers found in the literature

In the Figure 1 it is also possible to observe the number of studies that offer evidences for each barrier. The studies were conducted with different projects and different number of projects analyzed. In the following sections we will discuss the barriers found and the evidences that support the problem.

3.1 Social Interactions

This category represents the barriers related to the way newcomers interact with the community, including who are the members they exchange messages, the size of their contact network, how they communicate and how the community communicate with them.

Socializing with Project Members. The study conducted by [9] highlights the influence of social and political organization for newcomers willing to become a core developer. The author analyzes mailing list discussions and conduct an in-depth analysis of the socialization of a successful developer. He emphasizes the need to build an identity in the project: *"what the newcomer has to learn is how to participate and how to build an identity that will help get his ideas accepted and integrated."* Other authors also report the importance of socializing with central members. For example, Bird [2] quantitatively analyzed mailing lists and report that *"the social network measure, indegree, ... had a significant effect on immigration."*

All studies that analyzed the centrality/importance of the contacts found that the closer the newcomer is to the center of the community, the more successful the newcomer is. However, the newcomer usually does not choose who will answer her questions. So, when the most appropriate community members receive the newcomers, the chance of retention is higher.

Receiving (Timely and Proper) Response. The answers received from the community play an important role during newcomers onboarding. There are evidences of this barrier in seven studies found in this review. Some of them [11,24,22,19] report only the impact of receiving a (timely) response from the community as a barrier. Other researchers [16,18,20] also report the impact of the content of responses (properness).

One of the studies that analyze the impact of the answer contents found that *"almost all non-returning newcomers can be attributed to not receiving a response or receiving a condescending response"* [16]. Regarding the studies that analyze the timely response, [11] analyzed mailing list archives and found that *"nearly 80% of newbie posts received replies, and that receiving timely responses, especially within 48 hours, was positively correlated with future participation."*

We can see that community social skills can influence newcomers' decision to contribute to the project. Generally, newcomers demand attention and friendly hands to start contributing. We understand that core members need to stop their main tasks to receive newcomers with no guarantee that they will contribute. However, a good reception can be crucial to retain more newcomers.

Sending a Correct/Meaningful Message. [16] reported a problem related to newcomers' communication behavior. By analyzing the history of a support forum they found that *"the newcomers who used informative subject lines for their first message improved chances of getting responses as well as getting their problems solved by the community ... if the newcomer does not post comprehensible messages or uses a language that the forum responders do not understand...."* Therefore, newcomers who want to be welcomed by the community should focus on the quality of their community-oriented initial interactions.

Finding Mentorship/Expertise. Easiness to find an expert or a mentor is also evidenced in some studies [4,6]. Cubranic et al. [6] report that *"It can be* *difficult for newcomers to join such groups [OSS projects] because it is hard to obtain effective mentoring."* To alleviate it, Canfora et al. [4] proposed a tool that recommends mentors to newcomers. They evaluated the tool by surveying some project members and found that mentoring is important to newcomers.

Mentorship is presented as be a good way to help newcomers. However, its actual applicability need to be studied in deep. It is not clear if this kind of policy can be applied in OSS communities, as it depends on experienced volunteers to do this specific task.

3.2 Finding a Way to Start

This category represents the barriers related to difficulties that newcomers face when trying to find the right place to start contributing.

Finding Appropriate Issue/Task. Finding the appropriate task to work on was classified as a barrier. Park and Jensen [13] reported that *"... subjects expressed a need for information specific to newcomers, for instance, how to get involved and become active (e.g. communication channels, available sources of information for starters, etc.), what to contribute to (e.g. open issues, required features, sample tasks to start with), and working practices."*

Von Krogh et al. [22] also report on this issue. They found that *"in 56.7% of the cases members of the community encouraged the new participants to find some part of the software architecture to work on that would match with their specialized knowledge. In only 16.7% of the cases new participants were both encouraged to join and given specific technical tasks to work on."* This occurs because, according to their interviews, the community expects new participants to find their own task to work on instead of receiving a specific piece of work.

Communities point of view is that newcomers should be able to find the most appropriate task themselves, as reported by [22]. However, other researches show that the projects should give special attention to this issue [1,5,13].

Finding the Correct Artifacts to Fix an Issue. When the newcomer find a task to work on, another issue can impact his contribution: how to find the correct artifacts. Cubranic et al. [6] proposed Hipikat, a tool that recommends artifacts that are relevant to a task that a newcomer is trying to perform. When conducting an experiment with Eclipse project, they found that *"newcomers can use the information presented by Hipikat to achieve results comparable in quality and correctness to those of more experienced members of the team."*

Newcomers really need support on finding code artifacts related to the chosen task, as projects' structure/architecture are not always trivial and straightforward. So, projects would benefit from tools like Hipikat to support newcomers first steps, as evidenced by the study conducted by Cubranic et al. [6].

3.3 Code Issues

This category comprises the barriers that are related to the source code of the products. To contribute a newcomer usually needs to change existing source code. Therefore, it is necessary for the newcomers to have enough knowledge about the code to start their contributions.

Dealing with Code Complexity/Instability. Some studies focus on how code complexity can affect the newcomers to OSS projects. Studying SourceForge projects Midha et al. [12] show that *"cognitive [code] complexity has a strong negative influence on the number of contributions from new developers."* Stol et al. [20] highlight some complaints of newcomers about project structure/architecture of Open Source projects.

Understanding the Project Structure/Architecture. Stol et al. [20] highlight some complaints of newcomers about project structure of OSS projects. One subject reported that *"the hierarchy of the source code directory was counter intuitive for someone with little architecting experience."* Cubranic and Murphy [7] also present an issue faced during their experiment: *"We also had reports of a pair missing a relevant suggestion because they lacked knowledge about the overall structure of the system..."*

Park and Jensen [13] analyzed *"the potential benefits of information visualization in supporting newcomers through a controlled experiment."* They reported that *"providing visual information such as the class diagrams or dependency views ... would help new developers understand the structure of existing code and find problems to work on."*

The main complain regarding code is that its structure is hard to understand, and learning it would take too much time. The use of visualization [13], or even artifact recommendation tools [6] can alleviate this problem.

Setting up Local Workspace. The feedback obtained by Stol et al. [20] evidenced that newcomers have difficulties when setting up their environment. They reported some obstacles, for example: *"a challenge was that some [subjects] did not have any experience or knowledge on checking out source code from the version control system."* To welcome newcomers, the communities should provide easy access to tutorials and step-by-step cookbooks on how to obtain the code, setup up and build a local workspace.

3.4 Documentation Problems

Project documentation was also explored by some studies. Newcomers need to learn technical and social aspects of the project to contribute. Thus, problems related to documentation were recurrently reported.

Outdated Documentation. Steinmacher et al [19] report some issues faced by newcomers regarding outdated information: *"We can see many demotivating facts that occurred in this case: ... outdated information in the issue tracker made the developers waste time on an already existent feature and on checking each issue they pick to address..."* Stol et al.[20] also report issues regarding outdated documentation. They report that the subjects *"... were uncertain whether the available diagrams were still up to date and relevant for the current version of the software... Another reported challenge was the uncertainty whether the available documentation was up to date for the current version of the software."* Finding outdated documentation can make the newcomer gave up onboarding. So, documentation provided by projects must be up-to-date enough to support new developers.

Information Overload. [13,7] conduct experiments to assess the benefits that tools that support dealing with information overload can bring to newcomers. Cubranic et al.[7,6] presents a tool called Hipikat that aims at recommending source code artifacts that should be related to the issue a newcomer is working on. Park and Jensen [13] evaluate the use visualization tools to alleviate problems with overload and report that *"[the tools] provided more efficient ways to handle large amounts of data and understand dependencies in source code, reducing the learning curve and information overload experienced..."* A rich documentation is essential for newcomers trying to understand the projects. However, just providing a bunch of documentation leads to information overload. So, the project should provide easy ways to find this documentation.

Code Comments not Clear. In addition to outdated information and information overload, Stol et al. [20] report a problem related code comments. *"It was reported that the code was not very well documented, which made it more difficult to understand the source code."*

3.5 Newcomers' Previous Experience

This category comprises the barriers related to the experience of the newcomers regarding the project and the way they show this experience when joining the projects.

Lacking of Process and Practices. We found just one study presenting evidence of learning project practices as a barriers that can hinder the newcomers onboarding. The study conducted by Schilling et al. [15] found that previous knowledge regarding the project practices influence newcomer first steps. They report that *"familiarity with the coordination practices of the project team has a strong association with the time they spend on their projects after GSoC."*

Lacking of Domain Expertise. Von Krogh et al. [22] claim that *"feature gifts by newcomers emerge from the newcomers prior domain knowledge and user experience."* In the study conducted by Stol et al. [20], the subjects *"reported their unfamiliarity with the domain to be a hindrance."* So, newcomers who present previous domain knowledge have more chances to have a successful onboarding and to be well received by the community.

Lacking of Technical Expertise. Schilling et al. [15] reported that *"...level of practical development experience is strongly associated with their continued permanence."* Some studies report sending messages or patches to mailing list or issue tracker presenting previous technical skills can benefit the newcomer when joining. Stol et al. [20] evinced that *"when newcomers mentioned that they had already tried some options to fix their problem and have put efforts to look for a solution in the forums ...then the responders were quick to respond and were very helpful."*

Ducheneaut [9] reports that *"expertise is not enough to become a core member in Python: one also has to create material artifacts..."* Bird et al. [2] also investigate the impact of sending patches when start the contribution and found that *"demonstrated skill level via patch submission plays an important role in Python and Postgres."* All studies evidence that the newcomer who wish to contribute must check if the technical skills required for a given task or project match with their skills. Newcomers can be proactive and search for the required background, but the project must also provide ways for a newcomer to search which tasks fit to his technical profile.

3.6 Summary

Considering the model defined in the Figure 1, based upon barriers identified by using inductive coding through out the selected studies, we can summarize the evidences for each barrier as shown in Table 1. The category more throughly studied is social interaction, accounting for 13 studies while the others range from 8 to 9 related studies each.

Due to the nature of the approach to establish the model, there is at least one paper associated with a barrier. Considering the most studied one, we found that the most evidenced barriers are newcomers' previous technical experience, from Previous Experience category, and aspects regarding social network characteristics and response reception, from Social Interaction category.

Table 1. Studies that evidence each barrier

Category	Barrier	Studies
Finding a Way to Start	Finding appropriate task/issue	[1,5,13,22]
	Finding the correct artifacts to fix an issue	[6]
Newcomers' Previous Knowledge	Lacking of Domain expertise	[22,20]
	Lacking of Previous Technical Experience	[3,2,16,24,9,22,15]
	Lacking of Knowledge on processes and practices	[15]
Code Issues	Dealing with code complexity/instability	[3,12]
	Understanding architecture/code structure	[6,13,20]
	Setting up Local Workspace	[20]
Documentation Problems	Outdated documentation	[19,20]
	Code comments not clear	[20]
	Information overload	[6,13,20]
Social Interactions	Socializing with project members	[3,2,10,23,24,9,14]
	Receiving (timely and proper) response	[11,16,24,22,18,20,19]
	Sending a correct/meaningful message	[16,24]
	Finding Help - Mentor/Expert	[4,6,19]

4 Threats to Validity

This review may have missed some papers that address barriers encountered by newcomers to OSS projects, since we did not search into every possible source and some relevant papers may not contain the chosen terms. To reduce bias, we contacted some specialists in OSS domain. We adjusted the criteria to cover all relevant papers that were of our knowledge and conducted pilot studies.

Most part of the studies analyzed do not present as main focus analyzing the newcomers needs or the problems they face during their first steps. The papers that aim to analyze newcomers obstacles and problems focus on very specific problems. We know that it would be hard – or even impossible – to identify every problem that can affect newcomers. However, keeping the analysis to just some specific problems restricts the value of the outcomes and their applicability.

The findings of this review may have also been affected as the classification is a human process and it is based on some subjective criteria. In particular, the terms of the area do not have a common definition among all studies. The problems were classified based on inductive coding approach, which also relies on manual classification. To reduce interpretation bias related to human process, this review involved two researchers cross checking each paper for inclusion, and a third researcher responsible for reviewing and discussing the information generated after each step.

5 Conclusion

In this paper we identified 21 studies that evidence barriers that can hinder newcomers' onboarding in OSS projects. We aggregated the barriers evidenced across the literature in a single place. By using an inductive coding approach to organize the barriers, we proposed a model that relies on five categories: finding a way to start, social interactions, code issues, documentation problems and newcomers' knowledge. This model is the main contribution of this systematic review.

As a result of this classification we found that the most evidenced barriers are newcomers' lack of previous technical experience, receiving improper response from community, socializing with project members and finding the appropriate task/issue. This classification provides a baseline for further researches related to newcomers onboarding.

As future work we aim to conduct qualitative studies to confirm the barriers evidenced by the literature. We are conducting some interviews with OSS experienced developers and newcomers to verify what are the main barriers faced by newcomers from their perspective. We plan to refine the classification model based on the results of the interviews. Additionally, based on this model it is possible to propose strategies to offer a better support for newcomers, and study how these mechanisms can benefit newcomers.

Acknowledgements. The authors thank Fundacao Araucaria and CNPq (process 477831/2013-3) for the financial support. Marco Gerosa receives a grant from the CNPq and FAPESP. Igor Steinmacher receives grant from CAPES (BEX 2038-13-7).

References

1. Ben, X., Beijun, S., Weicheng, Y.: Mining developer contribution in open source software using visualization techniques. In: 3rd Intl. Conf. on Intelligent System Design and Engineering Applications (ISDEA), pp. 934–937 (2013)
2. Bird, C.: Sociotechnical coordination and collaboration in open source software. In: 2011 27th IEEE Intl. Conf. on Software Maintenance, pp. 568–573. IEEE CS (2011)
3. Bird, C., Gourley, A., Devanbu, P., Swaminathan, A., Hsu, G.: Open borders? immigration in open source projects. In: 4th Intl. Workshop on Mining Software Repositories, p. 6 (2007)
4. Canfora, G., Di Penta, M., Oliveto, R., Panichella, S.: Who is going to mentor newcomers in open source projects. In: Proceedings of the ACM SIGSOFT 20th Intl. Symposium on the Foundations of Soft. Eng., FSE 2012, Cary, NC (2012)
5. Capiluppi, A., Michlmayr, M.: From the cathedral to the bazaar: An empirical study of the lifecycle of volunteer community projects. In: Feller, J., Fitzgerald, B., Scacchi, W., Sillitti, A. (eds.) Open Source Development, Adoption and Innovation. IFIP, vol. 234, pp. 31–44. Springer, Heidelberg (2007)
6. Cubranic, D., Murphy, G., Singer, J., Booth, K.: Hipikat: a project memory for software development. IEEE Transactions on Soft. Eng. 31(6), 446–465 (2005)
7. Cubranic, D., Murphy, G.C.: Hipikat: recommending pertinent software development artifacts. In: 25th Intl. Conf. on Soft. Eng., pp. 408–418 (2003)
8. Dagenais, B., Ossher, H., Bellamy, R.K.E., Robillard, M.P., de Vries, J.P.: Moving into a new software project landscape. In: 32nd Intl. Conf. on Soft. Eng., vol. 1, pp. 275–284 (2010)
9. Ducheneaut, N.: Socialization in an open source software community: A sociotechnical analysis. Comput. Supported Coop. Work 14(4), 323–368 (2005)
10. He, P., Li, B., Huang, Y.: Applying centrality measures to the behavior analysis of developers in open source software community. In: 2nd Intl. Conf. on Cloud and Green Computing (CGC), pp. 418–423 (November 2012)
11. Jensen, C., King, S., Kuechler, V.: Joining free/open source software communities: An analysis of newbies' first interactions on project mailing lists. In: 44th Intl.I Intl. Conf. on System Sciences (HICSS), pp. 1–10 (January 2011)
12. Midha, V., Palvia, P., Singh, R., Kshetri, N.: Improving open source software maintenance. Journal of Computer Information Systems 50(3), 81–90 (2010)
13. Park, Y., Jensen, C.: Beyond pretty pictures: Examining the benefits of code visualization for open source newcomers. In: 5th Intl. Workshop on Visualizing Software for Understanding and Analysis, pp. 3–10 (September 2009)
14. Qureshi, I., Fang, Y.: Socialization in open source software projects: A growth mixture modeling approach. Org Res. Meth. 14(1), 208–238 (2011)
15. Schilling, A., Laumer, S., Weitzel, T.: Who will remain? an evaluation of actual person-job and person-team fit to predict developer retention in floss projects. In: 45th Intl. Conf. on System Sciences (HICSS), pp. 3446–3455. IEEE CS (2012)

16. Singh, V.: Newcomer integration and learning in technical support communities for open source software. In: 17th Intl. Conf. on Supporting Group Work, GROUP 2012, pp. 65–74. ACM (2012)
17. Steinmacher, I., Gerosa, M.A., Redmiles, D.: Attracting, onboarding, and retaining newcomer developers in open source software projects. In: Workshop: Global Software Development in a CSCW Perspective (2014)
18. Steinmacher, I., Wiese, I., Chaves, A.P., Gerosa, M.A.: Why do newcomers abandon open source software projects? In: Intl. Workshop on Coop. and Human Aspects of Soft. Eng., (CHASE) (June 2013)
19. Steinmacher, I., Wiese, I., Gerosa, M.A.: Recommending mentors to software project newcomers. In: 3rd Intl. Workshop on Recommendation Systems for Soft. Eng (RSSE), pp. 63–67 (June 2012)
20. Stol, K.-J., Avgeriou, P., Babar, M.A.: Identifying architectural patterns used in open source software: approaches and challenges. In: 14th Intl. Conf. on Evaluation and Assessment in Soft. Eng., Swinton, UK, pp. 91–100. BCS (2010)
21. Thomas, D.R.: A general inductive approach for analyzing qualitative evaluation data. American Journal of Evaluation 27(2), 237–246 (2006)
22. Von Krogh, G., Spaeth, S., Lakhani, K.R.: Community, joining, and specialization in open source software innovation: A case study. Res Policy 32(7), 1217–1241 (2003)
23. Zhou, M., Mockus, A.: Does the initial environment impact the future of developers. In: 33rd Intl. Conf. on Soft. Eng (ICSE), pp. 271–280 (May 2011)
24. Zhou, M., Mockus, A.: What make long term contributors: Willingness and opportunity in oss community. In: 34th Intl. Conf. on Soft. Eng (ICSE), pp. 518–528 (June 2012)

Does Contributor Characteristics Influence Future Participation? A Case Study on Google Chromium Issue Tracking System

Ayushi Rastogi and Ashish Sureka

Indraprastha Institute of Information Technology, Delhi
{ayushir,ashish}@iiitd.ac.in

Abstract. Understanding and measuring factors influencing future participation is relevant to organizations. This information is useful for planning and strategic decision-making. In this work, we measure contributor characteristics and compute attrition to investigate their relationship by mining Issue Tracking System. We conduct experiments on four year data extracted from Google Chromium Issue Tracking System. Experimental results show that the likelihood of future participation increases with increase in relevance of role in project and level of participation in previous time-interval.

1 Introduction

Contributors leaving project incurs significant direct and indirect costs [1] [2][4][5] thereby making it critical to retain existing contributors [6]. A data-driven approach to study future participation helps overcome challenges in existing practices by providing objectivity and transparency. In this work, we examine contributor characteristics namely role of participation and amount of work done as a measure of predicting future participation. We present an approach that uses statistical measures to classify contributors based on contribution into three mutually exclusive sets namely non-core team, loose core team and tight core team. Also we define attrition as a function of participation in two consecutive time intervals and study attrition rate for four roles (reporter, owner, commenter and cc'ed-contributor (cc'ed)) and three classes of contributors. We conduct experiments on Google Chromium Issue Tracking System (GC-ITS) dataset[1]. The dataset is extracted for four consecutive years and observations are recorded quarterly (3 months). In Issue Tracking Systems contributors play various roles. However, for this work we focus on four roles namely reporter, owner, commenter and cc'ed. These roles are associated with various stages of bug fixing lifecycle. Reporter reports the issue. Issue is fixed by owner in collaboration with commenters participating via threaded discussion forum. Owner may also request participation by cc'ing contributors. Contributors cc'ed are called for to serve specific request in issue.

---
[1] https://code.google.com/p/chromium

Table 1. Role based individual contribution pattern

Role	Mean	Std	Min	Max	.25Q	.5Q	.75Q	.9Q	.95Q	.99Q	Skew
Own	24.8	038.1	1	000559	3	10	32	67	94	172.1	003.8
Rep	02.9	008.8	1	000476	1	01	02	04	11	040.0	013.5
CC	20.0	051.4	1	001315	1	04	20	59	92	190.2	009.7
Com	14.0	422.9	1	119803	1	01	02	07	30	239.0	129.9

2 Empirical Analysis

Class of Contribution. Research shows that open source projects follow Pareto Distribution [3] that is 20% of contributors do 80% of work. However, our experimental results do not communicate the same. Table 1 shows that in GC-ITS contribution pattern is highly skewed for four roles. This observation demands statistical and data-driven approach to classify contribution of contributors. In Algorithm 1. we classify each contributor in one of the three mutually exclusive classes namely non-core team, loose core team and tight core team based on contribution. Non-Core Team (NCT) includes contributors who join project to address some specific issue they encountered. Loose Core Team (LCT) includes dedicated contributors with substantial contribution and Tight Core Team (TCT) includes contributors with relatively large contribution (with respect to LCT).

The input to the Algorithm 1. is contribution of contributors where each contributor plays at least one of the four roles. The output is contribution class of contributors calculated for all time-intervals. We measure the contribution for the role of owner, reporter and cc'ed as the total number of issues participated in time-interval defined quarterly. Similarly, for commenter we measure contribution in terms of total number of comments in time-interval. Further to ensure homogeneity for cross comparison we range normalize contribution in each role for all time-intervals on a scale of 1-100 (refer Equation 1). We then append the scores for four roles of contributor for a time-interval to create a structure. All missing values are assigned a negligibly small value (0.0001). We generate cumulative score that measures contribution in terms of relevance of role (weight of owner (W_O) > weight of reporter (W_R) > weight of cc'ed (W_{CC}) > weight of commenter (W_{Cm})) as W_O=0.5, W_R=0.25, W_{CC}=0.125 and W_{Cm}=0.125. The choice of weight depends on specific requirements and may vary for individuals. Assuming that participation in one role is independent of participation in other roles, we use weighted Geometric Mean to generate cumulative score (refer to Equation 2).The score generated ranges from 100 (highest) to approximately 0.0001 (negligible or no contribution). Next we find relative relevance of contribution that is the number of standard deviations datum is related to mean. We calculate Z-Score (refer to Equation 3). If value of Z is less than 0, it indicates that contributor is part of NCT. If value of Z is greater than 1 it defines TCT. Likewise value of Z greater than equal to 0 and less than equal to 1 implies LCT. We calculate contribution class for all time-intervals and return set of contributors for each class for each time-interval.

Algorithm 1. Algorithm to Identify Contribution Class

Require: struct{Owner O, Reporter R, CC'ed CC, Commenter Com} Contributor Con[]
Ensure: ConClass[][3]
1: **procedure** CONTRIBUTIONCLASS(Con)
2:     **for all** Time-Interval T_t **do**
3:        Normalize contribution using Range Normalization [1-100]

$$Y_i = \frac{(100-1) \times X_i}{Max(X) - Min(X)} \quad (1)$$

4:        Calculate weighted Geometric Mean where $w_O > w_R > w_{CC} >= w_{Com}$ and sum($w_O + w_R + w_{CC} + w_{Com}$) =1

$$Score(S) = O^{w_O} \times R^{w_R} \times CC^{w_{CC}} \times Com^{w_{Com}} \quad (2)$$

5:        Calculate Z-Score

$$Z = \frac{S - \mu}{\sigma} \quad (3)$$

6:        **if** $Z < 0$ **then**
7:           $ConClass_{T_t}[1] \leftarrow Con[Z< 0]$ ▷ Non-Core Team
8:        **else if** $Z > 1$ **then**
9:           $ConClass_{T_t}[2] \leftarrow Con[Z > 1]$ ▷ Tight Core Team
10:       **else**
11:          $ConClass_{T_t}[3] \leftarrow Con[Z >= 0 \; \&\& \; Z <= 1]$ ▷ Loose Core Team
12:       **end if**
13:       ConClass[T_t] $\leftarrow ConClass_{T_t}$
14:    **end for**
15:    **return** ConClass
16: **end procedure**

Attrition Rate. In this study, we believe that the contributor has left the project if duration of inactivity exceeds one time-interval (in this case measured quarterly). Thus Attrition Rate (AR) for time-interval T_t measures (in percentage) the fraction of contributors who left the project in time-interval T_t to the total number of contributors who participated in time-interval T_t and its preceding time-interval T_{t-1}.

Graphical Analysis of the Relationship between Contributor Characteristics and Future Participation. In Figure 1 horizontal axis of the plot represents consecutive time-intervals (measured quarterly) and vertical axis shows attrition rate. Colored lines (refer to legend) present attrition rate for contributors and their roles. We observe that contributor attrition rate (irrespective of roles as shown in black) fluctuates from 27% to 47%. Also we observe marked difference in attrition patterns for four roles. We see minimum attrition rate for owner (shown in blue) and maximum for reporter (shown in red). This follows the intuition that not every contributor can own issues. Figure 2 compares attrition rate of three classes of contributors namely non-core team, loose core team and tight core team (refer Algorithm 1.) across four years. We see in Figure 2 that the attrition rate for LCT and TCT ranges from 3% to 10% which is relatively less than the attrition rate for NCT (ranges between 27% and 43%). It indicates

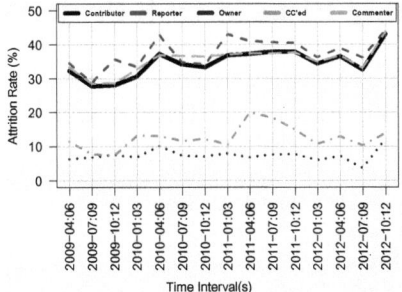 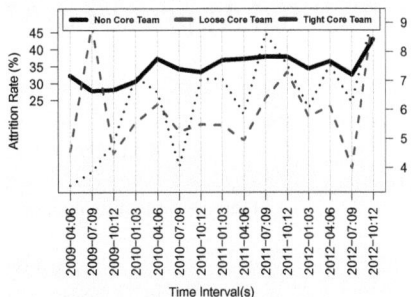

Fig. 1. Attrition rate of maintainers for four years

Fig. 2. Comparison of attrition rates for contribution classes for four years

that retention in project is directly related to degree of involvement in project. Also interestingly after initial fluctuations, attrition rate of TCT is higher than attrition rate of LCT indicating that TCT contributes relatively large however sporadically.

Acknowledgement. The work presented in this paper is supported by TCS Research Fellowship for PhD students awarded to the first author. The author would like to acknowledge Dr. Pamela Bhattacharya for useful insights and inputs.

References

1. Colazo, J., Fang, Y.: Impact of license choice on open source software development activity. Journal of the American Society for Information Science and Technology 60(5), 997–1011 (2009)
2. Izquierdo-Cortazar, D., Robles, G., Ortega, F., Gonzalez-Barahona, J.M.: Using software archaeology to measure knowledge loss in software projects due to developer turnover. In: HICSS 2009, pp. 1–10. IEEE (2009)
3. Robles, G., Gonzalez-Barahona, J.M.: Contributor turnover in libre software projects. In: Open Source Systems, pp. 273–286. Springer, Heidelberg (2006)
4. Robles, G., Gonzalez-Barahona, J.M., Herraiz, I.: Evolution of the core team of developers in libre software projects. In: MSR 2009. IEEE (2009)
5. Schilling, A., Laumer, S., Weitzel, T.: Together but apart: how spatial, temporal and cultural distances affect floss developers' project retention. In: Computers and People Research, pp. 167–172. ACM (2013)
6. Yu, Y., Benlian, A., Hess, T.: An empirical study of volunteer members' perceived turnover in open source software projects. In: HICSS 2012, pp. 3396–3405 (2012)

A Layered Approach to Managing Risks in OSS Projects

Xavier Franch[1], Ron Kenett[2], Fabio Mancinelli[3], Angelo Susi[4], David Ameller[1], Ron Ben-Jacob[2], and Alberto Siena[4]

[1] Universitat Politècnica de Catalunya (UPC), Barcelona, Spain
[2] KPA, Raanana, Israel
[3] XWiki, Paris, France
[4] Fondazione Bruno Kessler (FBK), Trento, Italy
{franch,dameller}@essi.upc.edu, {ron,ronb}@kpa-group.com,
fabio.mancinelli@xwiki.com, {susi,siena}@fbk.eu

Abstract. In this paper, we propose a layered approach to managing risks in OSS projects. We define three layers: the first one for defining risk drivers by collecting and summarising available data from different data sources, including human-provided contextual information; the second layer, for converting these risk drivers into risk indicators; the third layer for assessing how these indicators impact the business of the adopting organisation. The contributions are: 1) the complexity of gathering data is isolated in one layer using appropriate techniques, 2) the context needed to interpret this data is provided by expert involvement evaluating risk scenarios and answering questionnaires in a second layer, 3) a pattern-based approach and risk reasoning techniques to link risks to business goals is proposed in the third layer.

Keywords: OSS, Open Source, Risk Management, Layered Model.

1 Introduction

Translating dynamics of a complex system into focused management insights has been a challenge in various application domains [1]. In this paper we focus on organisations adopting, integrating and maintaining open source software (OSS) components in order to reduce time to market, introduce innovation and overcome development bottlenecks. Several companies have been observed to understand which are the main risks and risks indicators related to this OSS-related activities that are perceived by technical and business managers [2].

We propose a three layered approach: 1) the first layer focuses on collecting and summarising available data from different data sources, including human-provided contextual information; 2) the second layer, converts these data risk drivers into risk indicators [3] and 3) the third layer assesses how these indicators impact into the business of the adopting organisation. Key methodological as well as theoretical questions need to be answered to derive risk related insights from measurable data as described in [4][5] such as, the indicators to define for measuring risk events, how to operationalise an indicator into one or more specific metrics for measurement and the predictive ability of measurements related to risks events needs to be validated.

2 A Layered Approach to Risk Management

Here we describe the three layered approach to risk management in OSS projects that we are proposing in the EU project RISCOSS [6], see Fig. 1.

2.1 Layer 1: Raw Data and Risk Driver Measures

In this layer we deal with data collected from OSS communities and projects that determines the risk drivers. The data has a twofold nature. On the one side it refers to the characteristics of the OSS components developed by the communities, e.g. *Number of open bugs*, *Forum posts per day*, *Mails per day*, *Amount of documentation*. On the other hand, other measures highlight the structure of the community in terms of its evolution, e.g. changes in its roles and members and in the quality and quantity of relationships between them mainly via social network analysis techniques [7].

The data sources are community repositories, versioning systems, mail lists, bug trackers and forums, among others. Human intervention may be eventually needed because of: 1) data sources that may be unavailable for a particular component or community, 2) values that can eventually be calculated but require a dedicated activity to do so, 3) values that are not directly accessible or are very costly to compute.

2.2 Layer 2: Risk Indicators and Risk Model

In this layer we define the set of indicators of possible risks and models that allow linking these risks to the possible objectives of the adopting organisation. The indicators are variables extracted via the OSS community data analysis obtained from OSS project measurements and OSS community measurements as described before or via expert assessment. Several categories of indicators can be observed. Here we refer to three of them: 1) risk indicators related to OSS projects can be grouped following some criteria such as *Maintainability*; 2) risk indicators related to the communities coming from the aforementioned community measures, e.g. *Community activeness* or *Community cohesion*; 3) contextual risk indicators, elicited from experts, mainly depend on the objective of the organisation, e.g. *OSS business strategy*.

Here Statistical Analysis, Bayesian Networks and Social Network Analysis are exploited to determine values of risk. In particular:

Fig. 1. The 3-layered approach

- Statistical analysis of data from OSS communities allows determining the trends and distributions of data.
- Bayesian networks are used to link the community data gathered from the community data sources and the community risk metrics to the risk indicators and the community risk indicators using data generated by experts' assessment based on their experience in OSS adoption and community context.
- The community measures can be also analysed via Social Network Analysis techniques in order to understand the structure and evolution of the OSS community.

All the risk indicators will contribute to the definition of a risk model. This model allows the representation of the possible causes of risks, basically the risk indicators, and of their connection to the possible risk events for the adopter organisation. Moreover, the model also allows representing the impact that the possible risk events have on the strategic and business goals of the organisation.

2.3 Layer 3: Business Goals

Business goals describe which are the aims of the organization that adopts OSS. They are impacted by several kind of risks we summarise into four categories: 1) Strategic risks, mainly related to the company's strategy and plan, such as *Pricing Pressure*, *failures in comply Regulation*, *Industry or sector downturn*, or *Partner issues*; 2) Operational risks such as *poor capacity management* or *cost overrun*; 3) Financial risks such as *assets lost*, *debts* or *accounting problems*; 4) Hazard risks related to, for example, *macroeconomic conditions* or to *political issues*. Also in this case Bayesian networks may be used in order to link concepts from the two layers.

2.4 Modelling the Layers

Business goals are included in models that represent the ecosystem that blends together communities, OSS adopting organizations and other key actors. The key relationships between these actors are represented through dependencies in goal-oriented models expressed in the *i** language [8], which allow representing, and reasoning about, business goals and business processes. Reasoning is based upon different techniques, and in our layered context, we are particularly interested in bottom-up evaluation, since the leaves are directly linked to the risk model.

A typical model will include the two fundamental actors of the OSS ecosystems, the Community and the Adopter, and how they depend on each other; some of their internal goals and activities, and their further AND/OR decompositions. We have then the risk model with the risk event (e.g., *Risk of difficulty in code refinement*) that "impacts" one of the activities of the Adopter (e.g., *Bug Report*) and that is propa-gated up to the higher level activities. The risk event is identified via the measurement and statistical analysis of the behaviour of the community and on the expert intervention that can rate the evidence of a risk indicator via the Bayesian Networks.

3   Conclusions and Future Work

The RISCOSS framework is designed to face with the problem of risk management in OSS related projects in a holistic way, allowing to pass smoothly from the dimension of the measures to those related to the decision-making in contexts where several technical and business constraints are present.

We believe that the approach can give an effective way of overcoming problems related the adoption phase. In particular, the huge volume and potential heterogeneity of the data is isolated into a layer collecting the available and potential new techniques suited for this problem; the correct interpretation in the context of the adopting organization is made also with the help of experts that can evaluate specific scenarios of risks; a pattern-based approach and risk reasoning techniques is proposed in the third layer that can help in linking risks to business goals.

Several points have to be addressed in the following years of the project. We plan to refine the approach clearly defining the boundaries of the layers and adding to each one of the layers the suitable techniques for data reasoning. An important point here is that of developing the approach in such a way to be adapted to the needs of the particular organisation that should be able to also feed in a contextual way the necessary data to effectively exploit the approach. Also we plan to integrate better our results to those coming from projects with related aims, as FLOSSMetrics (http://flossmetrics.org/), QualiPSO project (http://qualipso.org/), QualOSS (http://www.qualoss.eu/) and OSSMETER (http://www.ossmeter.eu/).

Acknowledgments. This work is a result of the RISCOSS project, funded by the EC 7th Framework Programme FP7/2007-2013 under the agreement number 318249.

References

1. Harel, A., Kenett, R.S., Ruggeri, F.: Modeling Web Usability Diagnostics on the basis of Usage Statistics. In: Statistical Methods in eCommerce Research, Wiley (2009)
2. Li, J., Conradi, R., Slyngstad, O., Torchiano, M., Morisio, M., Bunse, C.: A State-of-the-Practice Survey of Risk Management in Development with Off-the-Shelf Software Components. IEEE Trans. Software Eng. 34(2) (2008)
3. Kenett, R.S., Baker, E.: Process Improvement and CMMI for Systems and Software: Planning, Implementation, and Management. Taylor and Francis, Auerbach Pub. (2010)
4. Ligaarden, O.S., Refsdal, A., Stolen, K.: ValidKI: A Method for Designing Key Indicators to Monitor the Fulfillment of Business Objectives. In: BUSTECH 2011 (2011)
5. Wallace, L., Keil, M.: Understanding software project risk: a cluster analysis. Inf. Manage. 42(1) (2004)
6. Franch, X., Susi, A., Annosi, M.C., Ayala, C., Glott, R., Gross, D., Kenett, R., Mancinelli, F., Ramsamy, P., Thomas, C., Ameller, D., Bannier, S., Nili Bergida, N., Blumenfeld, Y., Bouzereau, O., Costal, D., Dominguez, M., Haaland, K., López, L., Morandini, M., Siena, A.: Managing Risk in Open Source Software Adoption. In: ICSOFT 2013 (2013)
7. Salter-Townshend, M., White, A., Gollini, I., Murphy, T.B.: Review of statistical network analysis: models, algorithms and software. Statistical Analysis & Data Mining 5(4) (2012)
8. Yu, E.S.K.: Modelling strategic relationships for process reengineering. PhD thesis, University of Toronto, Toronto, Ont., Canada (1995)

A Methodology for Managing FOSS Migration Projects

Angel Goñi[1], Maheshwar Boodraj[2], and Yordanis Cabreja[1]

[1] Universidad de las Ciencias Informáticas (UCI),
Carretera a San Antonio de los Baños,
Km. 2 ½. Torrens, Municipio de La Lisa. La Habana, Cuba
{agoni,ycabreja}@uci.cu
[2] Mona School of Business and Management,
The University of the West Indies,
Mona, Kingston 7, Jamaica, W.I.
maheshwar.boodraj@uwimona.edu.jm

Abstract. Since 2005, the Free Software Center (CESOL) at the University of Information Science (UCI) in Havana, Cuba, has conducted several free and open source software (FOSS) migration projects for various organizations. The experience gained from these projects enabled the creation of a *FOSS Migration Methodology* which documented how the technical elements of a project of this kind should be executed. Despite the usefulness of this methodology, the projects that have been undertaken experienced difficulties that were, in most cases, directly related to their management. This research aims to improve the methodology and minimize management-related challenges thereby improving the quality of migration projects. The proposed methodology was applied in a project that ran in a higher education organization and the results prove that the methodology enhanced the quality of the migration project.

1 Introduction

The development of free and open source software (FOSS) and its advantages over the closed model that is prevalent today[1] have served as a catalyst for many organizations, including regional and national governments, to adopt FOSS. In 2005, the Cuban government decided to undertake the process of nationwide adoption of open source technologies. The implementation of different projects, in several Cuban organizations had made it possible to develop the *Cuban Migration Guide*[2], a document to govern and guide the process. This document was used in several public sector agencies to facilitate orderly and gradual migration. The number of projects executed[3] by the Free Software Center (CESOL) in Havana, Cuba, made possible the creation of a *FOSS Migration Methodology*[4] based on the most important guides released in Europe[5], Brazil[6], Peru[7], and Venezuela. It documented best practices and provided guidance for the team of specialists that would be involved in the project.

2   FOSS Migration Methodology

The *FOSS Migration Methodology* proposes several steps and workflows that meet the objective of efficient work organization and makes communication easier between members of the migration team and managers of the target organization. These stages and workflows can be seen in Figure 1.

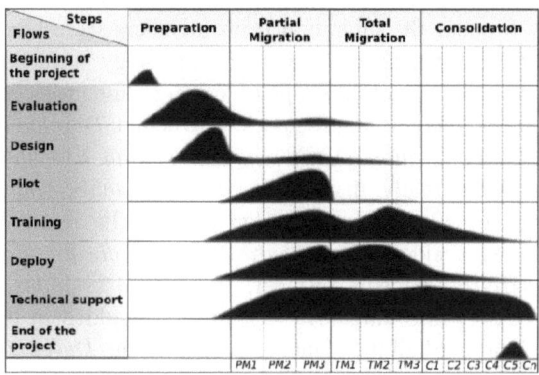

Fig. 1. Phases and workflows of the *FOSS Migration Methodology*

In 2012, several internal workshops were conducted at CESOL, in which team members shared their experiences regarding migration. The most common problems outlined were:

- Low executive commitment in the target organization;
- Unclear expectations of the target organization;
- Low staff participation in the target organization; and
- Late generation of project documentation.

These problems pointed to different areas of project management, which prompted a review of the migration methodology and project documentation, taking into account the following two critical project management challenges described in the literature[8-10]:

- Processes belonging to technology migration and training are properly identified while management processes are ignored; and
- Project management is performed as an isolated task and done experientially.

In several cases, the *Migration Plan* contained more technical details instead of focusing on planning elements. This created a voluminous document (over 100 pages), making it highly unreadable and limiting its usefulness for managers of the target organizations. The *FOSS Migration Methodology* and the migration guides[11] initially ignored the elements of project management, which necessitated the introduction of workflows (*Beginning of the project* and *End of the project*) to compensate for the lack of structured project management. To address the problems mentioned above, the *FOSS Migration Methodology* was improved by introducing project management elements that were originally overlooked. These elements were based on the *Project Management Body of Knowledge – Fourth Edition (PMBOK Guide)[9]*. During the improvement of this methodology, several changes were made to the *FOSS Migration Methodology* including the modified use of some of the processes proposed by *PMBOK Guide*. The migration project was split into two main phases: diagnostic phase and technology migration phase. The *Human Resource Management* and *Procurement Management* knowledge areas were removed in the final methodology because the migration team members were hired and trained by CESOL, and the purchase and acquisition of technology were done by the target organizations and were outside of the scope of the migration projects.

Several *Integration Management* processes were added: *Develop diagnostic report, Plan consultancy, Plan migration, Develop migration plan, Execute consultancy*, and *Execute migration*. Additionally, the *Develop project management plan* process was substituted with the *Develop migration plan* process. In *Scope Management*, the following processes were added: *Gather hardware and software information, Analyze project feasibility, Migrate network services, Migrate workstations*, and *Train users*. The initiating process *Create diagnostic schedule* was added in time management, along with the unification of all the planning processes into *Develop schedule*. Hence, the methodology adapts much of the project management processes from the *PMBOK Guide* to the particular characteristics of migration projects. It also omits unnecessary processes and adds others that complement the FOSS migration activities.

3 Results and Evaluation

In order to evaluate the impact of the improved *FOSS Migration Methodology*, the methodology was applied to a FOSS migration project of a higher education institution which had 202 staff members, 120 workstations, and 2 servers. The purpose of the FOSS migration was to improve project efficiency while attaining high levels of user satisfaction. The efficiency of the project was measured in duration (in days) of the two main phases of the migration: the diagnostic phase and the technology migration phase.

The diagnostic phase was completed in 16 days - almost half of the average duration recorded in the schedule of several previous projects. These results support the heavy redesigning of the processes involved in this phase and the introduction of the consultancy report as the main deliverable of this stage[11]. The technology migration phase was finished in 58 days, which was significantly less than that obtained in previous comparable undertakings. However, there are many external factors that could have affected these results and the authors cannot decisively ascertain how much of the reduction in time could be directly attributed to the improved *FOSS Migration Methodology*. Two surveys were conducted in order to measure users' satisfaction with the results of the migration project. One was administered to managers and the other to staff. Six managers of the entity were surveyed, representing 100% of senior management. Four of them were satisfied, one was neutral and another was dissatisfied with the migration results. In the case of staff, 30 teachers were surveyed, representing 55% of the academic staff involved in the migration. The results showed twenty satisfied, four neutral and four dissatisfied.

4 Conclusions

This research was conducted to improve the *FOSS Migration Methodology* to minimize management-related challenges and improve the quality of migration projects. The improved *FOSS Migration Methodology*, which adopts a process-based approach from the *PMBOK Guide*, was applied in a project that ran in a higher education organization and the results prove that the methodology enhanced the quality of the migration project. Specifically, the improved methodology provided increased efficiency resulting from shorter activity durations, higher levels of management and staff satisfaction, enhanced communication from more timely and comprehensive documentation, and improved project management guidance.

References

1. Raymond, E.S.: The Cathedral & the Bazaar, p. 258. O'Reilly Media, Sebastopol (2001)
2. Paumier, R., Pérez, Y.: Guía Cubana Para La Migración a Swl, La Habana, Cuba (2007)
3. Goñi, A.: Experiencias de la migración a NOVA del área docente de la facultad 10 de la UCI, La Habana, Cuba: Centro Coordinador para la Formación y el Desarrollo del Capital Humano del Ministerio de la Informática y las Comunicaciones (2009)
4. Pérez, Y., Méndez, J., Goñi, A.: Metodología cubana de migración a Código Abierto, La Habana, Cuba (2012)
5. Hnizdur, S.: Directrices IDA de migración a software de fuentes abiertas. European Communities, Surrey, United Kingdom (2003)
6. Brasileño, G., Libre, G.: Referencia de Migración para Software Libre del Gobierno Federal, G.d.T.M.p.S. Libre, Brasilia, Brasil (2004)
7. INEI, Guía para la Migración de Software Libre en las Entidades Públicas. Instituto Nacional de Estadística e Informática, Lima (2002)
8. CMMI, CMMI® for Services, Version 1.3. Software Engineering Institute, Pittsburgh (2010)
9. PMI, Guía de los Fundamentos para la Dirección de Proyectos. Project Management Institute, Inc., Pennsylvania (2008)
10. IPMA, ICB IPMA Competence Baseline Version 3, Nijkerk, The Netherlands (2006)
11. Goñi, A.: Metodología para la gestión de proyectos de Consultoría en Migración a Tecnologías de Software Libre y Código Abierto, p. 106. Universidad de las Ciencias Informáticas (2012)

The Agile Management of Development Projects of Software Combining Scrum, Kanban and Expert Consultation

Michel Evaristo Febles Parker[*] and Yusleydi Fernández del Monte

University of Informatics Sciences (UCI), Havana. Cuba, 537-8372531
{mfparker,ydelmonte}@uci.cu

Abstract. At the University of Informatics Sciences (UCI), Havana, Cuba, it is found The Center of Free Solutions of Software (CESOL) who has an informatic project named "Auditing of Source Code" (ACF). This project has as objective to develop an open source software solution to auditing the source code of several software solutions with an agile projects management. In the present investigation have been showed the experiences obtained in the mixed application of two methods of agile projects management; Kanban and Scrum, together with the method Judgment of Expert, during the stage of construction of the lifecycle of ACF, when it is was performed a quality auditing by specialists of the CALISOFT company. In the auditing were detected several errors and to resolve them was necessary to estimate efforts, time and to revalue the lifecycle of the project. Moreover, the investigation show how this method can be used as a guide for young project managers for a correct planification and how can be used as a personal organizational method.

Keywords: Scrum, Kanban, Agile management of projects.

1 Introduction

The Methods of agile software project management are guides for planning and control thereof. Currently several software development methodologies are focused on this style. The free software applications by the need to respond quickly to the constant changes in its requirements, technology and its short development period, are who most use them. In the Department of Operating System of the center CESOL, are developed open source software solutions using free tools and agile methodologies, organized in various development projects. Besides, the UCI is working with a view to improving the quality of his process of development up to Level 2 of CMMI. To check the correct execution of the model, audits and reviews are performed to the projects by the company specialists CALISOFT, the institution responsible for validating the quality of the process. ACF is a project that belong to the center CESOL, where a system to auditing the source code of software systems is made using an agile management and free software tools to develop. The present

---

[*] Corresponding author.

investigation shows the experiences gained during the combined application of the methods of agile project management, Kankan and Scrum, along with expert judgment method during the construction phase of the life cycle of the ACF project which was audited quality. Therefore the objective of this investigation is to show the experiences gained during the combined application of the agile methods project management, Scrum and Kankan, along with expert judgment, to achieve a pleasant management of a computer project.

2 Discussion

The agile management of projects is a management able to adapt and respond to new requirements and changes dictated by the environment [1] .Inside of this model are found the agile methodologies of development of software, that seek the early delivery of incremental software [2], among which are Extreme Programming (XP), Adaptive Software Development (ASD), Open Unified Process (OpenUp), Kanban and Scrum. Scrum is a process that apply regularly a set of best practices for working collaboratively. [3] Divide the team into small specialized groups managed by themselves, dividing the work into a list of small tasks or requirements for deliverables. Sort the list by priority and estimated the relative effort of each element [4]. Kanban is used to monitor the progress of work in the context of a production line and is currently used for agile project management, often with Scrum (known as Scrumban) [5].

CMMI is a model that is not focused on the principles of agile development. It is an adaptable guide to raise the quality level of the software development process of an entity. Particularly in the UCI there is a project called Programa de mejora, currently at version 3.4, in order to adapt CMMI to the different development centers that exist in it; lightening the documentation as possible, in order to fulfill model without making conflict with the environments of agile development. Particularly the case study is an example of a stage of the life cycle of the ACF project where was used the estimation method in view of solving the macro task "Fix no conformities identified during the audit quality", conformed by small subtasks.

2.1 How to Combine Both Methods?

It starts with a set of tasks or requirements (the term is to taste) to perform. Then it proceeds to prioritize tasks using various criteria defined by the person responsible for managing the process. These may be important for the customer, the level of complexity, the amount of resources required for the implementation and the dependency among the requirements. The criteria should not be less than three. Each task is evaluated using these criteria according to a metric that can also be defined by the person who manage the application of the method, preferably [1-5], [1-10] and [5-10]. After evaluating each task, the values obtained for each criterion are added together and this is the value to use as a criterion for prioritizing tasks, sorting them in descending leaving those with highest numerical value as the first to be executed. In the event that the comparison test match, you can optionally choose the order that those tasks will have between them. Later proceeds to define the time duration of the tasks. To make this process intervals are defined from the values obtained as a result of the comparison test; intervals can also be defined optionally. These intervals are associated durations for tasks. In the interval where the criterion of value is within, the time corresponding to the interval is associated to the task. The times are defined using analogies of old tasks, expert consultation, experience and personal judgment.

2.2 Experiences in the Combined Use of Both Methods

In June 2013 the ACF project was audited by specialists CALISOFT where a set of non-conformances that must be resolved in the shortest time as possible for the project were consistent with the quality model and resume its planning in the shortest possible time too. The initial group of tasks to be performed was as follows:

1-Perform document "Technical project", 2- Perform document "Project plan", 3- Perform document "Glossary of terms", 4-Perform document "Validation of the requirements", 5-Perform document "Plan of iteration", 6-Perform document "Work item list", 7-Perform document "Use case specification", 8-Perform document "Use case model, 9-Perform document "Specification of the requirements", 10-Perform document "Vision", 11-Perform document of architecture, 12-Perform document "Art state of the product to develop", 13-Perform document "List of risks" and 14-Perform document "Requirements of support".

The criteria defined for determining priority were complexity (task difficulty), size (effort needed to accomplish the task), importance (importance to the project) and the interest (interest of the project team of to execute the task), being the metric used 1-5. The result of the prioritization was the following list of tasks: Tasks 1, 2, 3, 4, 5 and 6 with priority 20; task 7 with priority 18; tasks 8, 9,10,11 and 12 with priority 17; task 13 priority 13 and task 14 with priority 12. The estimation of time intervals defined initially were: the tasks with priority [20] will last 3 days, with priority [19-18] will last 2 days and with priority [1-17] will last one day, which estimate a total of 29 days. Splitting the time between the number of workers on the project who is three, is obtained as a result approximately 9.7 days. To restrict the number of tasks was taken into account that two person on the team had the ability to perform two tasks simultaneously, for that reason was defined a working limit of 5 tasks for the columns Assigned, Developing and Reviewing.

Table 1. View of the Kanban board at the end of the first day of work, it can see that the task 3 was completed on the first day

Task list	Assigned	In development	In review	Finished	% of execution
8,9,10,11,12,13,14	2,5,6,7	1,4		3	7%

The second day, during a review of the remaining tasks was determined reassess the priority of task 9, leaving with a score of 20 therefore ascends to be the first task to execute.

Table 2. View of the Kanban board at the end of the second day of work

Task list	Assigned	In development	In review	Finished	% of execution
10,11,12,13,14	2,9,8	1,5,6,7		3,4	14%

At the end of the third day of work, the tasks 2 and 10 were assigned and the tasks 9 and 8 upgraded to review. An active risk1 in the ACF project is the lack material resources and eventually in the stage in question was materialized, affecting the implementation of task 11. Therefore, as corrective action was determined to eliminate the task 11 of the board because at that time do not had the necessary resources to execute it.

Table 3. View of the Kanban board at the end of the fourth day of work

Task list	Assigned	In development	In review	Finished	% of execution
11	2,12,13,14	1,5,6,7		3,4,9,8,10	35.7%

Table 4. View of the Kanban board at the end of the fifth day of work

Task list	Assigned	In development	In review	Finished	% of execution
	2	1,5,6		3,4,9,8,10,12,13,14,11,7	71.4%

At the end of the sixth day of work the task 2 entered into development and tasks 1, 5, 6 and 2 into review when the seventh day finished, on both days the percentage of implementation was 71.4%. On the eighth day all tasks were completed for a 100% of execution.

3 Lessons Learned

It is possible reassess the task duration by increasing or decreasing their priority based on the new that arises in the development of software. The feedback determine that for this case, the future tasks that are similar to the first 6 and number 9, the duration should be 8 days, which represents over five days than estimated. For the tasks of 2 days, must add them one day and the tasks of 1 day behaved as estimated, allowing it to update the initial estimates.

It can be estimated and reassign tasks at the moment, allowing time to mitigate the risks that may exist. During process execution task 9 was prioritized again, flexibly changing the allocation of the task by a need of the development team. Following the realization of a risk associated with the project, task 11 was removed as a corrective measure.

In the prioritizing the tasks influence the characteristics of the team in terms of their skills and behaviors. The meetings of checking allow analysis of the performance of tasks and receive feedback from the experiences of the entire team to make adjustments to improve the planning, estimation and execution.

Allows to analyze in short term, the trends of the development team and take steps to improve. Allows to know the speed of the work team and for inexperienced project leaders to estimate and to update that estimate at the same time that the project it is running.

It can work objectively by prioritizing tasks. The prioritization of the requirements allows to obtain a list of work focused on the key elements to achieve the project objectives; because each time a task is completed the final product evolves.

4 Conclusions

The investigation arrived at four conclusions. The first is that was showed the experiences gained during the joint implementation of Kanban and Scrum are detailed, along with the expert consultation to achieve a pleasant management in the project ACF. As a second conclusion, is that the combination of these methods allows a more precise estimate of the work, mainly for inexperienced project leaders and ordinary people. The third conclusion is that with his application the project can quickly reach his objectives that the product evolves in each review, that the planning be flexible and analyzes the existence of risks, his mitigation and that the entire work team participate in the management of the project. The fourth conclusion is that Scrum and Kanban complement themselves, by the characteristics of an agile environment there will always be changes in the requirements or the tasks during the development process, being necessary to insert them in that process. The board of Kanban by him selves, describes the workflow very well, join it with the prioritized list of SCRUM, the possibility of Kanban to modify the tasks without having to wait for the next iteration and a workflow guided by goals, demonstrate why is best to use them together.

References

1. Palacios, J.: Agile project management: basics concepts: conceptos básicos (2006), http://www.navegapolis.net/files/s/NST-003_01.pdf
2. Pressman, R.: Software engineering, 6th edn., ch. 4 Agile Development, p. 77. McGraw-Hill
3. Schwaber, K., Sutherland, J.: The Scrum Guide (2013), https://www.scrum.org/Portals/0/Documents/Scrum%20Guides/2013/Scrum-Guide-ES.pdf#zoom=100
4. Kniberg, H., Skarin, M.: Kanban and Scrum – getting the best of both (2010)
5. Garzás, J.: What is the Kanban method for project management (November 2011), https://eventioz.com.ar/e/el-metodo-kanban-creando-un-cambio-evolutivo-exito

An Exploration of Code Quality in FOSS Projects

Iftekhar Ahmed, Soroush Ghorashi, and Carlos Jensen

School of Electrical Engineering and Computer Science
Oregon State University, Corvallis, OR, USA
{ahmed,cjensen}@eecs.oregonstate.edu
ghorashs@onid.oregonstate.edu

Abstract. It is a widely held belief that Free/Open Source Software (FOSS) development leads to the creation of software with the same, if not higher quality compared to that created using proprietary software development models. However there is little research on evaluating the quality of FOSS code, and the impact of project characteristics such as age, number of core developers, code-base size, etc. In this exploratory study, we examined 110 FOSS projects, measuring the quality of the code and architectural design using code smells. We found that, contrary to our expectations, the overall quality of the code is not affected by the size of the code base, but that it was negatively impacted by the growth of the number of code contributors. Our results also show that projects with more core developers don't necessarily have better code quality.

Keywords: Code Quality, Success Metrics, FOSS, Open Source Software.

1 Introduction

Free/Open Source Software (FOSS) is associated with collaborative development model bringing developers together from different geographical locations to create cost effective and efficient software. The adoption of FOSS is growing; Walli et al. found that, out of the 512 U.S. companies surveyed, 87% of them used FOSS [43]. Another survey in 2007 also found that not only does FOSS have a significant market share, but that this development model in many cases produces the more reliable and highest performing software option [45].

The quality of FOSS software has been subject to debate. Though some studies have argued for that FOSS projects can produce high-quality code [33, 36, 6], critics often point to the lack of formal project management practices and requirements as roadblocks to reliably achieving high software quality [35]. That said, the FOSS community has developed its own methods of quality assurance and quality control [33], and researchers have shown that FOSS projects produce more secure code due to peer-review and the openness of the code [15]. Despite this controversy, there is a lack of large-scale empirical studies of objective code quality .

Evaluating the quality of software can be a difficult task because there are a large number of properties that could be evaluated [16, 44], such as functionality, adherence to specifications, security, usability, etc. [18, 6]. It is not clear that these factors, even when combined, would give an adequate definition of quality. As a result many researchers have used different project characteristics as a substitute measure of the quality of a project, including longevity [13], operational software characteristics [31], number of open bugs [5], etc. None of these properties measure code quality directly (e.g. open bug reports is at best an indirect measure, as it is confounded by the size of the code base and the quality and extent of testing). One could argue that in the absence of requirements and objective definitions of quality, the more objective measure to focus on would be the quality of the code itself.

We decided to examine the quality of FOSS source code using two types of code smells: implementation level code smells and design level code smells, two objective measures of code quality. A code smell is not an error or problem in itself, but rather a set of heuristics for identifying poor coding practices or code structure, often associated with poor maintainability, bugs, or inefficiencies [19]. This definition refocuses code quality to the maintainability and efficiency of the code.

We used existing code-smell detection tools to analyze implementation level code smells (focusing on violations of standard coding practices), and design level smells (focusing on higher-level issues associated with architectural decisions), and see if there were interactions between these and other project characteristics. In this paper we sought to answer the following questions:

- What impact do project characteristics such as longevity, size of code base and number of developers have on the low-level quality of the code?
- What impact do project characteristics such as longevity, size of code base and number of developers have on the quality of design-level decisions?

The outline of this paper is as follows. We review related work in FOSS development model. Next we describe our research methodology. Then we present our findings followed by the discussion, and finally outline future work and conclusion.

2 Background

Two of most important characteristics of the FOSS development model, for the purposes of this paper, are the distributed development process and the reliance on unpaid developers. These two factors can have an important effect on software quality because they lead to decentralization and difficulty compelling developers to take specific actions. That said, these characteristics are not indicative of a lack of quality; there are plenty of examples of outstanding FOSS projects, such as Linux, Apache, Gnome, Mozilla, etc.

Most FOSS research is focused on these successful and popular projects [27, 34], while ignoring the many smaller FOSS projects that fail to reach this level of quality, popularity, or success. While a core reason for failure can be a lack of interest or volunteers [40], some suggest that lack of management is a major cause [38].

One key to the success of FOSS is that source code availability allows faster evolution and higher quality because of parallel development and the theory that having more eyes on the code identifies more bugs quickly [32]. Because of this, as Stamelos et al. [39] found, mature FOSS code is generally of good quality.

Many metrics have been used to define the success of an information system (IS) such as a FOSS project. Delone and McLean's IS success model is perhaps the most cited [8, 9, 7]. In this model they identify six measures of success, with system quality being the most important measure for user satisfaction and adoption [22].

Looking beyond FOSS, Boehm et al. [3] and Gorton and Liu [14] among many software engineering researchers, have explored different measures for software quality; including completeness, usability, testability, maintainability, reliability, efficiency and etc. that are all relevant to our analysis.

Code smells are symptoms of problems in the source code, and an indicator of where refactoring is needed [11]. This in turn has been associated with errors [23] and code maintainability problems [11]. Researchers have come up with different types of code smells depending on their type of impact [25]. The identification of code smells is typically done during development, testing, and maintenance.

Many approaches have been proposed for code smell detection, such as metric based [20] and meta-model based [28]. Metric based measures show that code smells impact software quality [24]. Most of the code smell detection tools are based on metrics analysis [20]. This static analysis based approach has its drawbacks. Fowler and Beck claimed that *"No set of metrics rivals informed human intuition"* [11] when it comes to deciding whether an instance of a code smell should be refactored.

3 Methodology

Our goal was to analyze the impact of different project characteristics on the overall quality of code, measured using coding practice violations and design level code smells. We wanted to see if project characteristics such as longevity, number of core developers, the size of the developer community, the frequency of code changes, and size of the code base were correlated with the quality of the resulting code. This could shed valuable light on prevalent assumptions about the evolution of FOSS projects, and prescribed best practices.

We sought to perform an analysis of representative FOSS projects, but we also wanted to control for as many external factors as possible. One such factor was the programming language used. We decided to restrict our study to Java for two reasons: First, Java is one of the most popular languages (according to Github [12] and the Tiobe index [41]). Second, the number and robustness of Source Code Analysis (SCA) and code smell detection tools for Java compared to other languages.

For code smell detection we built on the work of Fontana et al [10] and selected the InFusion [17] toolset. To identify the standard coding practice violations, we used "Codepro Analytix" [4] which uses a set of 643 rules collected from textbooks of java such as *The Elements of Java Style* [42] and *Effective java* [2].

We used the 110 most popular Java projects hosted on Github [12]. Popularity was measured using the number of stars given to projects, a product of the number of users who liked the project and chose to get updates about it. Core developers in this paper are defined as the group of contributors that contribute major portions of the commits and also act as reviewers for patches submitted by others. Table 1 gives an overview of the key characteristics of the projects in our dataset.

Table 1. Summary of the project characteristics

	Min	Max	Median	Average	Stddev
File count	480	462,846	22,811	51,270.0	74,354.5
LOC	701	968,287	45,564	10,702.1	175,146.2
# of commits	20	120,947	555	2,880.7	11,891.5
Age (days)	5	5,485	910	1,086.0	905.8
# of core developers	1	44	3	6.1	7.6
# of contributors	1	225	25	43.7	47.2
Popularity (stars)	695	7,252	1,011	1,376.5	987.8
Total design code smells	0	4,981	28	231.7	645.3
Total coding violations	18	4,894	921	1,074.1	908.9

In the second phase we analyzed the projects to check if project characteristics contributed towards specific design issues. We categorized code smells into broader categories, as suggested by the code smell literature [25]. Our categories were: Bloater, Object oriented abusers, Coupler, Dispensables, Encapsulators and Others.

- *Bloaters* are smells that leads the code to balloon so that it cannot be effectively handled. This includes data clumps, large class, long method, long parameter list and primitive obsession [25].
- *Object oriented abusers* are smells that do not fully exploit the advantages of object-oriented design. Some of the smells include Switch statements, parallel inheritance hierarchies, and alternative classes with different interfaces [29].
- The *Coupler* category contains smells related to high coupling between objects, in defiance of good object oriented design principles. Smells in this category include feature envy and inappropriate intimacy [29].
- The *Dispensable* category contains smells such as the lazy class, data class and duplicate codes [29].
- The *Encapsulators* category contains smells that deal with data communication or encapsulation, and includes message chain and middleman smells [29].
- *Others* is an aptly named catch-all category.

Two researchers independently performed this categorization. Inter-rater agreement was calculated using Cohen's Kappa. As the categorization was straightforward and there were only 6 categories, we achieved a Cohen-Kappa score of 0.90 after the first round of coding. According to Landis and Koch [19], Cohen-Kappa score greater than .85 indicates almost perfect agreement between coders.

4 Results

We calculated correlation coefficient between the number of low-level smells and the project characteristics mentioned earlier. None of these showed a Pearson Correlation Coefficient greater than 0.50, meaning that there was no strong linear correlation between any of the single properties and overall low-level code quality.

Next we used linear regression analysis with the intercept β_0 set to zero (absence of files or other variables results in zero code violations). Our model discarded Popularity and Number of commits as significant factors (see Table 2).

Table 2. Quality indicator of the linear model for measuring low-level code smells

Title	Estimate	Std. Error	t value
File count	-0.4401	0.1996	-2.205
LOC	0.3233	0.1549	2.087
# of core developers	-0.6321	0.2715	-2.328
# of contributors	0.8408	0.2020	4.162
Age (Days)	0.4563	0.1506	3.030
*R-Squared	**0.4735**		

To answer our second research question we checked for correlations between the number of design level smells and the project characteristics. Surprisingly we found that there was a strong liner relationship between the number of design level code smells and the number of core developers (Pearson Correlation Coefficient = 0.73).

Next we used linear regression analysis with the intercept β_0 set to zero (absence of files or other variables results in zero code violations). Our model discarded LOC, Number of contributors and Popularity as significant factors (see Table 3).

Table 3. Quality indicator of the linear model for measuring design level code smells

Title	Estimate	Std. Error	t value
File count	0.14303	0.10169	1.406
# of core developers	0.49993	0.08675	5.763
# of commits	0.16490	0.7962	2.071
Age (Days)	-0.13961	0.08079	-1.728
# of lines deleted	0.13450	0.07667	1.754
*R-Squared	**0.5349**		

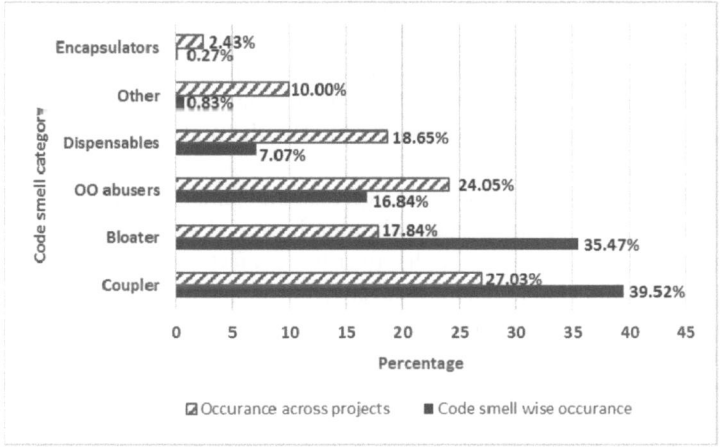

Fig. 1. Percentage occurrence of code smell categories

Figure 1 shows the mapping of design-level code smells into the six high-level categories discussed in our methodology section. We found that Couplers and Bloaters are the most common design-oriented code smells (solid bars), with 39.52% and 35.47% of all occurrences respectively. However, when we look at the percentage of projects that have at least one of these smells we see a much more even distribution (striped bars).

5 Discussion

The linear regression models that we found have R-squared values of 0.4735 and 0.5349 for low-level smells and design-oriented smells respectively. This means that a handful of project factors can account for roughly 50% of the variance in the sample. We also found that for the design smells, there was a strong correlation between the number of core developers and smells.

Looking at the data we see few interesting things standing out. For low level smells we were not surprised to see a correlation between the size of the code base or the number of contributors on one hand, and the number of smells on the other. This was expected; the more complex the code and the more coders there are, the harder it is to curate the code effectively, thus allowing more code smells to manifest. We also expected file count and number of core developers to have a positive effect on code quality, as the first leads to better organization, and the second better curating and supervision.

What was somewhat surprising was to see that the age of the code-base had a negative effect on code quality. One would expect the code to improve over time, but it appears as if this is not necessarily the case. It is likely that the urge to add new features, or the turnover of programmers outweigh any refactoring and fine-tuning performed by the community.

For high-level, or design oriented smells, we found that file count, number of commits and number of core developers actually led to an increase in the number of smells. Most of these correlations make sense. While the number of files help manage the low-level complexities of the code, adding files makes it harder to maintain the high-level conceptual design. Likewise, while adding core developers help projects stay on top of low-level code reviews, but hinders the high-level design vision. Too many cooks do seem to spoil he broth.

Unexpectedly we found that age did decrease the number of design smells, which means that though age is associated with more low-level smells, it is also associated with fewer design smells. It may be that over time, refactoring is more often aimed at removing design flaws rather than low-level coding convention violations.

Turning to the last part of our analysis, we find support for our interpretation of the results; the code smell categorization gives us a deeper understanding of the underlying causes for code smells. Coupler was the most common code smell, and represents the smells that causes high coupling between objects [29]. This might be an indication of a lack of adherence to object oriented design principles. We can hypothesize that, this type of ad hoc change happens due to the lack of knowledge about the design decisions of the project and due to low adoption rate of modeling and design tools in FOSS community [33].

Bloaters were the second most common code smell, and represent smells that lead the code to grow out of control, and is often caused by small changes and additions to the code [29]. Bloating is associated with centralized control structures using object-oriented languages, and Arisholm et al. identified that novice developers perform better with centralized control styles [1]. So it's most likely that novice developers turned contributors are pushing these changes. This potentially raises a question about quality assurance in FOSS.

In our analysis we have used the total number of coding standard violations and code smells as the indicator of FOSS projects quality. Identified regression models indicate that the number of core developers is correlated with quality of the FOSS project. This is in line with the findings of Sen et al., who found that number of developers reflect the "healthiness" of a FOSS project [37]. Contrary to one of the findings of Sen et al. we didn't find any strong relationship between the popularity of a project and the quality of the project. This can be explained by the "lurking" phenomenon that is prevalent in online communities [30] up to 90% [26]. The "lurkers" are members of the community, who do not participate in any activity except watching, so it make sense that they have little impact on the actual quality of the code. Further empirical analysis is required to identify the actual contribution of lurking towards this observation to make any concluding remarks on this topic.

6 Limitations

It's always difficult to generalize a diverse movement such as FOSS. While we tried to ensure diversity amongst the projects we selected, there were limitations to our methodology and selection criteria. We list the most important here:

Our analysis was of 110 FOSS projects, which though a reasonable sample, is small compared to a population of 375,486 projects. Though we selected the projects based on popularity we did not consider the application domain of the selected projects. During the data collection, we also did not exclude folders that were not directly related to the functionality of the system itself (test folders, contribution folders, demo folders, etc.).

7 Conclusion

We were able to show that there is a correlation between a number of important factors such as the longevity, the number of developers, and code-base size on one side, and the number of low-level code smells on the other. This implies that as projects grows and ages, the quality of the code decreases, unless counteracted by a larger group of core contributors to curate submissions. However, we also found that increasing the number of core developers negatively affects the higher-level design of the code. It is therefore important to carefully balance the size of the core group.

Acknowledgements. We would like to thank the Oregon State University HCI group for their input and feedback on the research.

References

1. Arisholm, E., Sjoberg, D.I.: Evaluating the effect of a delegated versus centralized control style on the maintainability of object-oriented software. IEEE Transactions on Software Engineering 30(8), 521–534 (2004)
2. Bloch, J.: Effective java. Addison-Wesley Professional (2008)
3. Boehm, B.W., Brown, J.R., Lipow, M.: Quantitative evaluation of software quality. In: Proc. 2nd International Conference on Software Engineering, pp. 592–605 (1976)
4. CodePro Analytix, https://developers.google.com/java-dev-tools/codepro/doc/
5. De Groot, A., Kügler, S., Adams, P.J., Gousios, G.: Call for quality: Open source software quality observation. In: Damiani, E., Fitzgerald, B., Scacchi, W., Scotto, M., Succi, G. (eds.) Proc.Open Source Systems. IFIP, pp. 57–62. Springer, Boston (2006)
6. del Bianco, V., Lavazza, L., Morasca, S., Taibi, D., Tosi, D.: An Investigation of the Users' Perception of OSS Quality. In: Ågerfalk, P., Boldyreff, C., González-Barahona, J.M., Madey, G.R., Noll, J. (eds.) OSS 2010. IFIP Advances in Information and Communication Technology, vol. 319, pp. 15–28. Springer, Heidelberg (2010)
7. DeLone, W.H., McLean, E.R.: Information Systems Success Revisited. In: Proc. of the 35th Hawaii International Conference on System Sciences (2002)
8. DeLone, W.H., McLean, E.R.: The DeLone and McLean Model of Information Systems Success: A Ten-Year Update. Journal of Management Information Systems, 9–30 (2003)
9. DeLone, W.H., McLean, E.R.: Information Systems Success: The Quest for the Dependent Variable. Information Systems Research, 60–95
10. Fontana, F.A., Mariani, E., Morniroli, A., Sormani, R., Tonello, A.: An experience report on using code smells detection tools. In: Software Testing, Verification and Validation Workshops (ICSTW), pp. 450–457 (2011)
11. Fowler, M.: Refactoring: improving the design of existing code. Addison-Wesley Professional (1999)
12. Github, https://github.com
13. Golden, B.: Making Open Source Ready for the Enterprise, The Open Source Maturity Model. Extracted From Succeeding with Open Source. Addison-Wesley Publishing Company (2005)
14. Gorton, I., Liu, A.: Software Component Quality Assessment in Practice: Successes and Practical Impediments. In: Proc. of the 24th International Conference on Software Engineering, pp. 555–558. IEEE Computer Society
15. Hoepman, J.H., Jacobs, B.: Increased security through open source. Communications of the ACM 50(1), 79–83 (2007)
16. Sommerville, I.: Software Engineering. Pearson Education Limited, Essex (2001)
17. InFusion, http://www.intooitus.com/inFusion.html
18. Jung, H.W., Kim, S.G., Chung, C.S.: Measuring software product quality: A survey of ISO/IEC 9126. IEEE Software 21(5), 88–92 (2004)
19. Landis, J.R., Koch, G.G.: The measurement of observer agreement for categorical data. Biometrics 33, 159–174 (1977)
20. Lanza, M., Marinescu, R.: Object-oriented metrics in practice: using software metrics to characterize, evaluate, and improve the design of object-oriented systems. Springer (2006)

21. Lavazza, L., Morasca, S., Taibi, D., Tosi, D.: Predicting OSS trustworthiness on the basis of elementary code assessment. In: Proc. of ACM-IEEE International Symposium on Empirical Software Engineering and Measurement, p. 36 (2010)
22. Lee, S.Y.T., Kim, H.W., Gupta, S.: Measuring open source software success. Proc. Omega 37(2), 426–438 (2009)
23. Li, W., Shatnawi, R.: An empirical study of the bad smells and class error probability in the post-release object-oriented system evolution. Journal of Systems and Software 80(7), 1120–1128 (2007)
24. Marinescu, R.: Detecting design flaws via metrics in object-oriented systems. In: 39th International Conference and Exhibition on Technology of Object-Oriented Languages and Systems, TOOLS 39, pp. 173–182 (2001)
25. Marticorena, R., López, C., Crespo, Y.: Extending a taxonomy of bad code smells with metrics. In: Proc.7th ECCOP International Workshop on Object-Oriented Reengineering (WOOR), p. 6 (2006)
26. Mason, B.I.: Issues in virtual ethnography. In: Proc. of Ethnographic Studies in Real and Virtual Environments: Inhabited Information Spaces and Connected Communities, pp. 61–69. Edinburgh (1999)
27. Mockus, A., Fielding, R.T., Herbsleb, J.D.: Two case studies of open source software development: Apache and Mozilla. ACM Transactions on Software Engineering and Methodology 11, 309–346 (2002)
28. Moha, N., Rezgui, J., Guéhéneuc, Y.-G., Valtchev, P., El Boussaidi, G.: Using FCA to suggest refactorings to correct design defects. In: Yahia, S.B., Nguifo, E.M., Belohlavek, R. (eds.) CLA 2006. LNCS (LNAI), vol. 4923, pp. 269–275. Springer, Heidelberg (2008)
29. Mäntylä, M.: Bad smells in software-a taxonomy and an empirical study. Helsinki University of Technology (2003)
30. Nonnecke, B., Preece, J.: Lurker Demographics: Counting the Silent. In: Proc. CHI 2000, pp. 73–80 (2000)
31. Rating, B.R.: Business readiness rating for open source, http://openbrr.org
32. Raymond, E.: The cathedral and the bazaar. Knowledge, Technology & Policy 12(3), 23–49 (1999)
33. Robbins, J.: Adopting open source software engineering (OSSE) practices by adopting OSSE tools. In: Perspectives on Free and Open Source Software, pp. 245–264 (2005)
34. Koch, S., Schneider, G.: Effort, cooperation and coordination in an open source software project: GNOME. Information Systems Journal 12(1), 27–42 (2002)
35. Scacchi, W., Feller, J., Fitzgerald, B., Hissam, S., Lakhani, K.: Understanding free/open source software development processes. Software Process. Improvement and Practice 11(2), 95–105 (2006)
36. Schmidt, D.C., Porter, A.: Leveraging open-source communities to improve the quality & performance of open-source software. In: Proc. of the 1st Workshop on Open Source Software Engineering (2001)
37. Sen, R., Subramaniam, C., Nelson, M.L.: Open source software licenses: Strong-copyleft, non-copyleft, or somewhere in between? Decision Support Systems 52(1), 199–206 (2011)
38. Senyard, A., Michlmayr, M.: How to have a successful free software project. In: Proceedings of the 11th Asia-Pacific Software Engineering Conference, pp. 84–91. IEEE Computer Society, Busan (2004)
39. Stamelos, I., Angelis, L., Oikonomou, A., Bleris, G.L.: Code quality analysis in open source software development. Information Systems Journal 12(1), 43–60 (2002)
40. Subramaniam, C.: Determinants of open source software project success: A longitudinal study. Decision Support Systems 46, 576–585 (2009)

41. Tiobe, http://www.tiobe.com/index.php/content/paperinfo/tpci/index.html
42. Vermeulen, A. (ed.): The Elements of Java (TM) Style, vol. 15. Cambridge University Press (2000)
43. Walli, S., Gynn, D., Rotz, V.: The Growth of Open Source Software in Organization (2005), http://dirkriehle.com/wp-content/uploads/2008/03/wp_optaros_oss_usage_in_organizations.pdf
44. Wennergren, D.M.: Clarifying Guidance Regarding Open Source Software, OSS (2009), http://dodcio.defense.gov/Portals/0/Documents/FOSS/2009OSS.pdf
45. Wheeler, D.: Why Open Source Software/Free Software (OSS/FS,FOSS, or FLOSS)? Look at the Numbers (2007), http://www.dwheeler.com/oss_fs_why.html

Polytrix: A Pacto-Powered Polyglot Test Matrix

Max Lincoln and Fernando Alves

ThoughtWorks Brazil
Av. Gov. Agamenon Magalhães, 4779, 12° andar
Empresarial Isaac Newton
Ilha do Leite, Recife - PE
50070 160 - Brazil

Abstract. We have created a polyglot test framework named Polytrix to compare, benchmark, and independently verify a suite of open-source OpenStack SDKs that each target a different programming language. The framework validates sample code from each SDK against a shared test scenario to validate that each SDK correctly implements a given scenario. It uses Pacto for integration contract testing between the SDKs and the OpenStack services, and generates test reports that help compare and document each SDK. It is designed so interactive training materials can be generated in future versions.

1 Introduction

OpenStack was founded by Rackspace Hosting and NASA in July 2011 as an open-source system for building public and private clouds[1]. Since then, it has grown into a global open-source initiative, with contributions from more than 1800 engineers and 168 companies[2], and users in more than 132 countries[3]. Rackspace has continued to be an major contributor to OpenStack throughout the latest (Havana) release[4], and is also a major contributor to many projects that have grown around OpenStack. These other projects include OpenStack SDKs. These SDKs fall into two groups[5]:

> OpenStack SDKs: only support clouds built with OpenStack (including OpenStack private clouds as well as public clouds from Rackspace, HP, and IBM)

> Multi-cloud SDKs: support OpenStack in addition to other clouds

OpenStack has not officially selected supported SDKs, but Rackspace supports OpenStack SDKs written in .net (openstack.net), php (php-opencloud), python (pyrax) and go (gophercloud) as well as multi-cloud SDKs written in java (Apache jclouds), node.js (pkgcloud) and ruby (fog)[6]. The three multi-cloud SDKs are each governed by a different organization.

The OpenStack wiki does make the requirements of an SDK clear:

1. A set of language bindings that provide a language-level API for accessing OpenStack in a manner consistent with language standards.
2. A Getting Started document that shows how to use the API to access OpenStack powered clouds.

3. Detailed API reference documentation.
4. Tested sample code that you can use as a "starter kit" for your own OpenStack applications.
5. SDKs treat OpenStack as a blackbox and only interact with the REST/HTTP API.
6. Must be sustainable.
7. License must be compatible with Apache License v2.

The quality and completeness of SDKs can be difficult to assess, especially as the number of SDKs grows. The OpenStack wiki states: "What constitutes an official OpenStack SDK has not been determined. This is an area the needs more work."

This paper outlines a test framework and documentation generator codenamed Polytrix (Polyglot Testing Matrix) that assesses SDKs and produces documentation that may help them achieve officially supported status.

2 Solution Overview

Polytrix[7] was using Ruby's RSpec[8] testing language, the Pacto[9] integration contract testing library, and the Docco[10] documentation generator. Since we are testing many different languages, we needed a generic way to update dependencies before running tests. We used the GitHub 'script/bootstrap' pattern[11] to keep things language agnostic. Once each project was bootstrapped, the Polytrix invokes each test. We run a Pacto server in the background to act as a test stub, mock or spy[12], capturing and validating calls made to services against their expectations that are defined in files referred to as "Pacto contracts".

Figure 1 shows the flow of a Polytrix test. We first launch Pacto as a server so it can accept requests over HTTP from the SDKs (step 1). Then we locate (step 2) and execute (step 3) the code sample for each SDK. The test environment is configured so calls from the SDKs to OpenStack are proxied through Pacto (step 4). Pacto will validate each request and forward it to the real service (step 5), or could return a stubbed response for fast feedback. Validation includes checking the structure of each request and response against json-schema[13] defined in the "Pacto Contract", as well as checking for expected HTTP headers. Once the sample code finishes executing, we check that it executed without errors (step 6). We then compare our test expectations against the actual interactions observed by Pacto, making sure that the expected services were called with appropriate data (step 7). If necessary, we could query OpenStack to check the final state (step 8), but step 7 is usually sufficient. Finally, we instruct Pacto to teardown any resources created during testing by sending deletion requests to counteract any creation requests that were detected during the test (step 8).

The role that Pacto fills is similar to that of the CC helper used by Codecademy's Submission Correctness Tests. The primary difference is that the CC helper validates locally invoked functions, why Pacto validates the invocation of remote services. This similarity makes it possible to enhance the test suite to work as an interactive online programming course for the SDK, similar to how CS Circles[14] or Codecademy teach general programming.

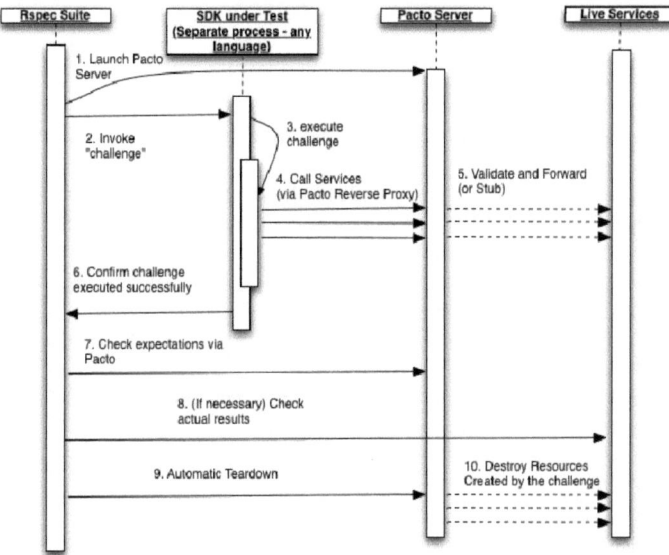

Fig. 1. Polytrix Testing Sequence

Once the test is complete, we generate documentation from the test metadata and extract additional documentation from the code samples using Docco[10], a tool inspired by Literate Programming[15].

3 Tutorial and Future Work

In the tutorial we will demonstrate how Polytrix can help projects like OpenStack assess multiple SDKs by:

- Collecting and displaying sample code for common scenarios in a central location so that the completeness and consistency with language standards of language-level bindings can be reviewed.
- Testing code samples, including verifying that they interact with as expect with REST/HTTP APIs.
- Generating Getting Started Guides and Starter Kit documentation from the test suite.

We will also discuss how Polytrix can be adapted to validate user input instead of pre-written code samples in order to create an interactive online programming course for the SDK.

References

1. 'rCloud' Is Your Cloud: The OpenStack Journey - The Official Rackspace Blog. Rackspace Hosting, http://www.rackspace.com/blog/rcloud-is-your-cloud-the-openstack-journey/

2. Stackalytics | OpenStack community contribution in all releases, `http://stackalytics.com/?release=all` (last modified January 31, 2014, accessed January 31, 2014)
3. Home » OpenStack Open Source Cloud Computing Software, `http://www.openstack.org/` (last modified January 31, 2014, accessed January 31, 2014)
4. Who Built Havana. OpenStack Summit (2013)(November 8, 2013) (print)
5. SDKs - OpenStack, `https://wiki.openstack.org/wiki/SDKs` (last modified January 31, 2014, accessed January 31, 2014)
6. Rackspace Developer Center, `http://developer.rackspace.com/` (last modified February 02, 2014, accessed February 02, 2014)
7. rackerlabs/polytrix, `https://github.com/rackerlabs/polytrix` (last modified February 04, 2014)
8. Chelimsky, D.: The RSpec book: behaviour-driven development with RSpec, Cucumber, and Friends, Pragmatic, Lewisville, Tex (2010)
9. thoughtworks/pacto, `https://github.com/thoughtworks/pacto` (last modified February 03, 2014)
10. jashkenas/docco, `https://github.com/jashkenas/docco` (last modified February 04, 2014)
11. Bootstrapping consistency, `http://wynnnetherland.com/linked/2013012801/bootstrapping-consistency` (last modified January 28, 2014)
12. Meszaros, G.: xUnit Test Patterns: Refactoring Test Code. Pearson Education, vol. 1 (2007) ISBN 9780132797467
13. IETF, JSON Schema: core definitions and terminology, draft (2013), `http://tools.ietf.org/html/draft-zyp-json-schema-04`
14. Pritchard, D., Vasiga, T.: CS Circles: An In-Browser Python Course for Beginners. arXiv:1209.2166 [cs] (2012)
15. Knuth, D.E.: Literate Programming. Comput. J. 27, 97–111 (1984)

Flow Research SXP Agile Methodology for FOSS Projects

Gladys Marsi Peñalver Romero, Lisandra Isabel Leyva Samada,
and Abel Meneses Abad

University of Informatics Sciences (UCI), Roar of San Antonio de los Baños.
Km 2 1/2. Tor-rens. Havana, Cuba
gmpenalver@uci.cu

Abstract. This paper aims to explain a procedure that takes into account the different research processes carried out in developing an open-source, allowing control and management. This study is the SXP methodology applied in this type of project was carried out, allowing the validity of the basis of this research.

Keywords: methodology SXP, open-source, production, research, software.

1 Introduction

University of Informatics Sciences (UCI) has six faculties and several development centers which in turn are composed of productive projects. Each center specializes in a different line of development, and of them the Free Software Center (CESOL) is dedicated to the creation of the Cuban operating system (Department of Operating Systems (OS)) and migration processes lead to open source applications, from a model of integration of training, research and postgraduate (Department of Comprehensive Immigration Services, Counseling and Support (SIMAYS)).The process of migrating to open source companies is one of the most important services in addition to counseling, consulting, training and support offered, and to expedite the work of specialists have developed applications that automate many of the processes governing the service provided, raising the quality and quick response customer service. For those products SXP agile methodology created to develop projects with more speed and quality expected by the end user applies.

The great need for each project was implemented in less than one year, with frequent deliveries, with teams of no fewer than 10 members, where self-management and the ability of each of the members of the development team were some of the reasons why this methodology is used during the life cycle of these projects.

Although its application has fulfilled its main objective, did not include artifacts, and activities that define the research process that occurs when developing software, thus resulting in the need to integrate him a workflow process to collect research in the various projects, defining artifacts, activities and roles in order to obtain higher quality products, competitive, and I could control and disseminate knowledge that occurs during software development.

2 Development

When starting a project already has how to carry out the development process, although at the beginning an entire previous research, which identifies the object of study, the current domestic and international situation is realized, which is framed in the first phase of any project; and it is the same theory that is followed in SXP.

2.1 Development Methodology SXP

SXP is a Cuban hybrid agile [1], which is premised on avoiding duplication of efforts and customer integration into the development team which ensures no need for extensive documentation, and thus is well recorded which will be used in a future reuse. Behold the good practices of the agile methodologies XP and SCRUM, besides the quality guidelines defined by Calisoft, which is the entity responsible for monitoring the status of each of the projects that are developed in the centers of development of the UCI and model CMMi quality. It is divided into four phases which form the basis of the structure of your project file, these are: Planning, Definition, Development, Delivery and Maintenance. Each of these phases is made up of a number of activities which are generating artifacts documenting the process and guide the development of the products.

2.2 ¿Where the Investigative Work Reflected SXP?

In the Planning-Definition phase is where the vision is established, the expectations are set and securing project financing is done. However not considered a research-oriented approach, because it is not spoken any time of writing the research project or research tasks that are thought to develop as a result of the production process, although this element should lead to torque software development process. When starting the development cycle, where the implementation of the product is made, also carried hand research, which can be very intense according to the different technical aspects to be analyzed in order to define the most sensible when developing. A final product is obtained nothing but the study documented, being within the knowledge of the researcher. Sometimes it happens that other projects need this information when developing, and not have them accessible, they should start their research from scratch, which impacts the product development time. So we can conclude that given the number of projects being developed, despite being guided by the methodology SXP lose the opportunity to document the wealth of research that systematically develop them. Which is knowledge that is generated and that in turn runs the risk of being lost with time, showing scientific activity affected by this situation. As it became necessary to develop a research stream with the inquiry process that takes place during the production of the software, allowing it to monitor and disseminate the knowledge produced.

2.3 Flow Research with Artifacts and SXP Role for Methodology

The methodology has a workflow that contains a number of artifacts that enables the control and management of research. This research stream will not be located in a specific phase, because its location will be chosen by the working group for each of the projects. Although it is recommended that some of the artifacts begin construction in early stage (Planning - Definition). Below are by each of the evidence gathering activity that takes place in the workflow:

Artifact Research Development Plan (IDP) is a document that reflected the initial da-planning to develop investigative activities, and should be done by the Research Manager, taking into account the characteristics defined for this role.

Artifact State of the Art, is a deliverable that will be developed after a preliminary investigation, so it is proposed that has its beginnings in the planning phase - definition, which does not mean they can not come changes in the remaining phases. This device may involve different roles of the development team in its preparation should not only be developed by the Research Manager.

Artifact Research Report suggests that develops in the development stage and that's where you are getting the results of investigations carried out. This device may involve different roles of the development team in its preparation should not only be developed by the Research Manager.

Role: Research Manager

- Person responsible to manage all the research tasks that develop.
- It is responsible for planning the development of research, verify compliance and quality of them.
- It is responsible for the preparation of PDI.
- It need not have computer skills, but if some domain of Research Methodology.

2.4 Valoración de la Propuesta

This method is novel because there is no documentation on the pervasive culture of research conducted during the development of software.

Although some productive research projects managed without defining roles artifacts or take responsibility for the control and dissemination of the same. It must be emphasized that this research stream can be implemented not only by the SXP methodology, but may be included in any other software methodologies analyzed in this research considering the stage of development that is more convenient when incorporate it. The artifacts can be applied to any research task running on a productive project because it meets the adjustable parameters to their characteristics. With the implementation of this proposal fails to meet four key fundamentals: First, the problem of the organization of research in productive projects is solved, and second, the basis for publication are encouraged, in addition to the socialization of knowledge produced in software development, third, a favorable economic impact is obtained in projects, reusing the basis of research to accelerate the development of future products and finally, the training of human resources for software production is favored.

3 Conclusions

In general we can draw the following conclusions:

Insertion flow enables research documenting the research tasks in the production of FOSS projects, it includes new artifacts and roles to software development methodology analyzed (SXP).

The knowledge gained in studies for re- use in new developments is guaranteed, and that serve as a basis for specialists who are interested in such projects.

Controls the research tasks in productive projects, raising the engagement of specialists to exchange knowledge and scientific-technical work.

It encourages collaborative work between developers and members of development teams, because centralizing research in a repository can be accessed

PDI, State of the Art and Research Report: Quality control and documentation of the investigative work done, with the inclusion of artifacts is guaranteed.

References

1. Romero, P., Marsi, G.: Metodología ágil para proyectos de software libre, SXP (2008)
2. Calderín, A., Yenin, I.: Procedimiento para el control de tareas investigativas en la producción de software en la UCI. Pág.61 Panorama IT
3. Raycel, C.F., Susel, G.P.: Propuesta de un expediente, para los proyectos productivos del Polo de Software Libre, de la Facultad 10. Pág. 34 (2008)
4. Galán, F.J., Cañete, J.M.: Qué se Entiende en España por Investigación en Ingeniería del Software (2005)
5. Genova, G., Llorens, J., Nuviola, J.: Métodos abductivos en Ingeniería del Software (2005)
6. Galán, G.F.J., Miguel, C.J.: Qué se Entiende, en España, por Investigación en Ingeniería del Software?
7. Rafael, M.: Metodología de desarrollo de software (2010)

How to Support Newcomers Onboarding to Open Source Software Projects

Igor Steinmacher[1] and Marco Aurélio Gerosa[2]

[1] DACOM – UTFPR Campo Mourao – PR – Brazil
`igorfs@utfpr.edu.br`
[2] IME – USP Sao Paulo – SP - Brazil
`gerosa@ime.usp.br`

Abstract. While onboarding an open source software (OSS) project, contributors face many different barriers that hinder their contribution, leading in many cases to dropout. Many projects leverage the contribution of outsiders and the sustainability of the project relies on retaining some of these newcomers. In this research, we aim at understanding the barriers that hinder onboarding of newcomers to OSS projects, by means of different empirical approaches, and proposing a set of strategies that can be used to support the first step of newcomers.

1   Introduction

Open Source Software (OSS) communities are generally self-organized and dynamic, counting with contributions of volunteers spread all over the globe. These communities demand a high influx of newcomers to keep alive [1]. According to Qureshi and Fang [2], motivate, engage and retain developers is a way to foster a sustainable amount of developers in a project. However, newcomers face difficulties and obstacles when they start interacting to the project, resulting in a high dropout rate [5]. Newcomers need to learn both social and technical aspects of the project, exploring mailing lists, wikis, issue trackers and source code repositories [3].

In a previous study [6], some developers who tried to onboard in two well-known OSS projects reported the obstacles they faced. Developers indicated that the lack of awareness and guidance during the course of their first steps (setup and choosing the right mean to start) discouraged further contributions. Therefore, a major challenge for OSS projects is to provide ways to support the onboarding of newcomers.

The goal of this research is to explore the barriers faced by newcomers to onboard to OSS projects and analyzing which strategies can be used to support newcomers to overcome these barriers. The overall question to be answer in this thesis is: *"How to support the onboarding of newcomers in OSS projects?"* To guide answering this question, some specific research questions were defined for this thesis:

- What are the barriers faced by newcomers when onboarding OSS projects?
- What are the forces that influence the joining and retention of developers in OSS projects?
- What strategies can be used to help newcomers?

2 Research Design

The research of this thesis uses different empirical methods to answer the research questions defined. This research is a result of the combination of systematic literature reviews, interviews, case study and experiments. The research design is composed of three phases and some studies, as presented in Figure 1, described below:

- **Warm Up.** This phase consists of one case study (S1) conducted in order to motivate the problem addressed by this thesis and to investigate whether absence of response, politeness, usefulness or the author of answers influence the onboarding of newcomers in an OSS project.
- **Phase I – Find Barriers.** This phase is composed of three studies that objective to raise what are the barriers that hinder newcomers onboarding to OSS projects. We will empirically gather the barriers from the literature (S2), by means of Systematic Literature Review, and from the OSS community (S3), using interviews with project members and feedbacks from newcomers. The results obtained in the systematic review and in the qualitative analysis will be confirmed by means of a survey to be administered with OSS community members (S4).
- **Phase II – Proposing Support to Newcomer Onboarding.** In this phase, the goal is to consolidate the barriers that hinder newcomers onboarding to OSS and map them to proposed strategies that can support newcomers overcoming them. We plan to propose solutions based on: (i) suggestions and studies that emerged during Phase I; (ii) systematic review on awareness mechanisms (S5); and (iii) state-of-the-practice, by analyzing the strategies that some projects currently adopt to support newcomers.

Based on the results of the survey conducted during Phase I, we will select a subset with the most relevant barriers to further explore in this phase. We plan to design and conduct a controlled experiment to assess the effectiveness of using the proposed solutions to the selected barriers (S6).

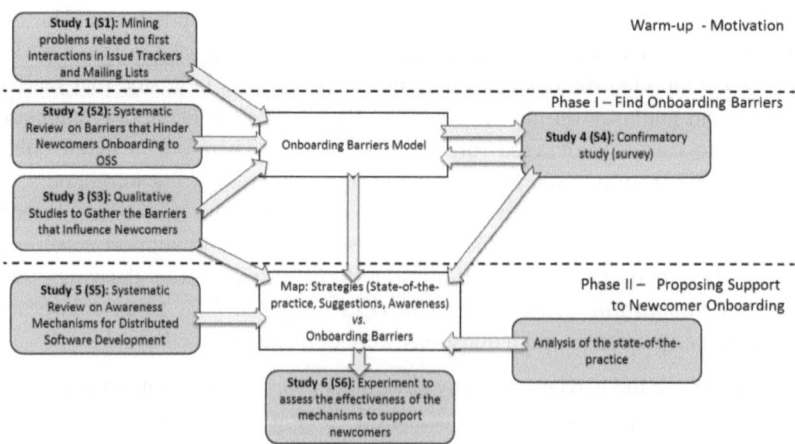

Fig. 1. Research Design

3   Preliminary Results

In order to situate our problem and delimit the scope of this thesis, we analyzed the existent literature and proposed a model [4] that represents the stages (onboarding and contributing) that are common to and the forces that are influential to newcomers being drawn or pushed away from a project.

We conducted a case study [5] to analyze newcomers dropout reasons. We collected five years of communication of the list of emails from developers and discussions from the task manager (Jira). There was no indication that the number of responses influences the withdrawal. We found evidence that receiving inadequate answers affects the decision of newcomers to abandon the project

We also collected feedback from some developers that tried to contribute to two OSS projects [6]. The developers reported some demotivating facts: unanswered emails, outdated documentation, outdated issues that resulted in waste of time and flaming threads.

We conducted a systematic literature review aiming at identifying the barriers faced by newcomers to OSS projects. As a result, we provided a hierarchical model that relies on five categories: social interactions, newcomers' previous knowledge, finding a way to start, documentation problems, and code issues. Some barriers that hinder newcomers to OSS were also identified using a qualitative analysis on data obtained from newcomers and members of OSS projects. The results enabled us to create a model composed of 38 problems, grouped into seven different categories.

Acknowledgements. The authors thank CAPES (BEX 2038-13-7) and CNPq (process 477831/2013-3) for the financial support. Marco Gerosa receives a grant from the CNPq and FAPESP.

References

[1] Park, Y., Jensen, C.: Beyond pretty pictures: Examining the benefits of code visualization for Open Source newcomers. In: 5th Intl. Workshop on Visualizing Software for Understanding and Analysis, pp. 3–10 (2009)
[2] Qureshi, I., Fang, Y.: Socialization in Open Source Software Projects: A Growth Mixture Modeling Approach. Org. Res. Methods. 14(1), 208–238 (2011)
[3] Scacchi, W.: Understanding the requirements for developing open source software systems. Software, IEE Proceedings 149(1), 24–39 (2002)
[4] Steinmacher, I., Gerosa, M.A., Redmiles, D.: Attracting, Onboarding, and Retaining Newcomer Developers in Open Source Software Projects. In: Workshop on Global Software Development in a CSCW Perspective (2014)
[5] Steinmacher, I., Wiese, I., Chaves, A.P., Gerosa, M.A.: Why do newcomers abandon open source software projects? In: Intl. Workshop on Cooperative and Human Aspects of Software Engineering (CHASE), pp. 25–32 (2013)
[6] Steinmacher, I., Wiese, I.S., Gerosa, M.A.: In: 2012 Third Intl. Workshop on Recommendation Systems for Software Engineering (RSSE), pp. 63–67 (2012)

The Census of the Brazilian Open-Source Community

Gustavo Pinto[1] and Fernando Kamei[1,2]

[1] Federal University of Pernambuco, Recife, PE, Brazil
ghlp@cin.ufpe.br
[2] Federal Institute of Sertão Pernambucano, Ouricuri, PE, Brazil
fernando.kenji@ifsertao-pe.edu.br

Abstract. During a long time, software engineering research has been trying to better understand open-source communities and uncover two fundamental questions: (i) *who* are the contributors and (i) *why* they contribute. Most of these researches focus on well-known OSS projects, but little is known about the OSS movement in emerging countries. In this paper, we attempt to fill this gap by presenting a picture of the Brazilian open-source contributor. To achieve this goal, we examined activities from more than 12,400 programmers on Github, during the period of a year. Subsequently, we correlate our findings with a survey that was answered by more than 1,000 active contributors. Our results show that exists an OSS trend in Brazil: most part of the contributors are active, performing around 30 contributions per year, and they contribute to OSS basically by altruism.

Keywords: OSS, Github, Brazilian OSS Community.

1 Introduction

The idea of Open Source Software (OSS) has gained more and more attention in the last years. OSS is usually developed by an internet-based community of programmers, without necessarily being paid by an institution. These programmers often rely on code hosting websites to share their contributions. Nowadays, a number of code hosting websites are available, such as SourceForge and Github. A unique characteristic of these websites is that they provide a collaborative environment with a high degree of social transparency. This makes programmers' contributions much more visible and traceable. In Github, for example, one can easily access programmers' personal information through a public interface, which otherwise it might be difficult, or even impossible, to find elsewhere.

Over the last decade, some researchers have studied open-source communities and contributions [5,13,14], although only few of them are regarding the open-source community in those websites [3,9]. Hitherto, however, there is a lack in the literature of comprehensive studies targeting the Brazilian open-source community. By the beginning of the 21th century, with a population size of 200 millions, Brazil is South America's most influential country, an economic giant and one of the world's biggest democracies. Moreover, Brazil has also been a hotbed of open source activities in recent years. Government agencies, private industry and universities have been teaching and implementing open source solutions, creating local centers of knowledge and gain expertise around open source in the country.

Nonetheless, besides the huge open-source investments made in Brazil, few is known about the Brazilian open-source contributor. We argue that it is an important question because, for example, if there is a lack of open-source programmers in one region, the government could create better incentives for software companies in there. On the other hand, if there is a huge number of experienced contributors, software companies could open more opportunities in that location. In this paper we conducted a comprehensive study to understand *who* is this contributor, and *why* (s)he contributes to OSS. To achieve this goal, we extracted data from more than 12,000 Brazilian users on Github. In addiction, we conducted a survey comprising only the active contributors, that is: the contributor that has performed at least one commit in a year. With this data, we are able to uncover these three main research questions:

RQ1: Who is the Brazilian open-source contributor? We found out that most of the Brazilian open-source contributor are male, have between 20-30 years, are currently enrolled in an undergrad course, have between 2 to 5 years of professional software experience, and 2 to 5 years of OSS contribution. Most of them perform around 30 commits per year, but 20.35% of the contributors perform 80.32% of the contributions. Also, we observed that they are basically formed by hobbyist, instead of programmers being paid by software companies.

RQ2: Do the Brazilian contributions to open-source increase over the time? We have found out an existing open-source trend in Brazil. We noticed an increment of over 15% in the absolute number of contributions in one year. However, we also observed that these contributions are not related to more work done by the same users. In fact, the great number of contributors performing few contributions is the reason of this increment.

RQ3: Why do Brazilian programmers contribute to OSS? We found out more than forty motivational factors that motivate users to contribute to OSS. The most common were to **help the community** and **to improve the software that they use** with 19.40% and 17.75%, respectively. Moreover, the majority of the Brazilian OSS community are not motivated by self-marketing but by altruism.

2 Study Design

In this section we present our research questions and our research approach.

2.1 Research Questions

The goal of the study is to better understand the Brazilian open-source community. For this purpose, we elicited three research questions.

- **RQ1:** Who is the Brazilian open-source contributor?
- **RQ2:** Do the Brazilian open-source contributions increase over the time?
- **RQ3:** Why do Brazilian programmers contribute to OSS?

In order to address the research questions, we conducted a two-phase research, adopting a sequential mixed-method approach. In Phase 1, we collected data about Brazilian users on Github (see Section 2.2). After that, on Phase 2, we performed a survey targeting only the active users (see Section 2.3).

2.2 Phase 1 – Mining Github

We used Github data as provided through the GHTorrent project [1], an off-line mirror of the data offered through the Github API. Up to September 2013, more than 2 million users and 5 millions projects were collected. We then have performed a query on the GHTorrent database, searching through the name of the 26 Brazilian capitals and their states. We also considered the capital of the country. Thus, we have searched users in 53 locations.

After this process, we found a total of 12,485 users. After that, we then have used the Github API to gather more information about these users, and we observed that almost 1,000 of these users have closed their accounts. Also, we manually removed false-positive users, that is, users that location name is similar to the Brazilian ones, but actually the location is located in a different country. Therefore, the population of this study consist of 11,411 Github users. Our data comprise the period of October 2012 to September 2013.

2.3 Phase 2 – Survey

The questionnaire used in this work was based on the recommendations of Kitchenham et al. [7], following the phases prescribed by the authors: planning, creating the questionnaire, defining the target audience, evaluating, conducting the survey, and analyzing the results. The questionnaire has 10 questions and was structured to limit responses to multiple-choice, Likert scales (responses given in a scale), and also free-forms. After defining all questions on the questionnaire, we obtained feedback iteratively and clarified and rephrased some questions and explanations. This feedback was obtained from analysis and discussion with a group of specialists and also from one pilot of the survey. Together with the instructions of the questionnaire, we included some simple examples as an attempt to clarify our intent.

Our target population is formed by programmers that have performed at least one commit to an open-source software during the analyzed period. After we identified these users, we sent them an email inviting to participate. Over a period of 20 days, we obtained 1,039 responses, resulting in a 16.68% of response rate. This response rate is more than trice higher than the response rates found in software engineering surveys [6]. The complete list of questions and their answers are available at the companion website[2].

---

[1] http://ghtorrent.org/
[2] http://bit.ly/Brazilian_OSS

3 Study Results

This section presents our results organized by the research questions.

3.1 Who Is the Brazilian Open-Source Contributor?

We started with a population of 11,411 Github users. However, not all all of them are active: during the period of Oct 2012 – Sep 2013, the majority part of the users (6,228 or 54.57% of them) have performed at least one commit. We consider these users as our active group. Hereafter, the analysis of this study will only encompass this group.

The Brazilian open-source community is also more active than the overall Github community (23.38% of the Github users are active). Still, our survey data show that this number of active users was not expected. On average, the respondents believe that only 12% of the population is active (SD: 23.12) and most of them (37.34%) believe that the contributors perform only 5 to 20 contributions per year. Moreover, our data show that most of the respondents are male (97.48% of them), have between 20 and 30 years (66.92%), and are currently enrolled in an undergrad course (36.36%). The participants experience in professional software development is mostly between 2 to 5 years (28.86%). The open-source experience is mostly between 2 to 5 years (36.36%), and then between 1 and 2 years (35.58%). On average, the respondents have contributed to open-source using three different programming languages (3rd Quartile = 5, max = 14). We then analyzed the correlation between the age and number of programming languages used using Pearson Correlation [8]. For Pearson $|r| < 0.3$ indicates small correlation, $0.3 > |r| < 0.5$ indicates medium correlation, and $|r| > 0.5$ indicates strong correlation. We then found out a small correlation ($r = 0.05702044$), which suggests that the age is not related to the number of programming languages used. On the other hand, we found out a strong correlation between age and professional experience ($r = 0.5636623$) and a medium correlation between age and OSS experience ($r = 0.4440954$), which is fairly similar to the results found by Morrison et al. [12].

An important property is the programming language used. Most of the contributions were done using JavaScript (15.88%), PHP (12%), C (11.58%), Java (11.04%), Python (10.43%), Ruby (9.52%), ShellScript (7.40%) and C++ (6.17%). Others languages have 16.60% of contributions. Most of the contributions are done using dynamic languages (60.04% of them) and also object oriented languages (45.81%) instead of functional languages (2.19%) or mixed programming languages (21.07%). One reason to this is because most of the Brazilian undergrad courses still rely on well-known programming languages, such as C and Java, in their basic courses. Also, Brazil has a number of strong open-source communities. For example, we have one of the largest Java User Group of the world, and one of the most active JavaScript, PHP and Ruby communities in Latin America. As reported elsewhere [2], open-source communities play an important role in the education of novice programmers.

Moreover, the number of contributions per user is generally less than 30 per year (3rd Quartile = 50, max = 4,795, mean = 14, Stand. Dev. = 177). The reason of this huge standard deviation is due to the high number of outliers (11.70% of the total). The outliers are consisted by users that have performed more than 120 contributions in one year. The Figure 1 shows the distribution of the contributions per user over the year.

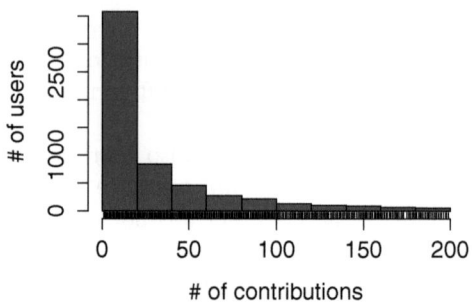

Fig. 1. The number of contributions per user over a year. We have limited the number contributions up to 200 to increase the readability of the figure.

As we can see, most of the active users are low actives (65.59%), performing less than 20 commits per year. On the other hand, we have observed that the Pareto's Laws fits perfectly in our data: 20.35% of the active users perform 80.32% of the contributions. Furthermore, the number of projects contributed per user is generally less than 5 (95% percentile: 12, 90% percentile: 6, 80% percentile: 3), with a median of 3. We have also observed that they spent on average less than 1 day per month working on open-source projects. Our data show that 8.73% of them do contributions every month, and only 2.47% of them do contributions every week. With this data, we can assume that the Brazilian open-source contributor is basically formed by hobbyist, instead of programmers being paid by software companies.

Finally, we mapped the user based on their regions. Brazil is divided into five geopolitical regions. The **North** includes seven states and it includes the Brazilian part of the Amazon rainforest. It is sparsely populated, and its economy is based on plant and mineral exploration. The **Northeast** includes nine states, an arid climate, and an economy based on agriculture, mainly sugarcane. The population is concentrated in a few large cities in the coast. The **Midwest** includes three states and Brasilia, the capital of Brazil. Sparsely populated, its economy is based on large farms agriculture and livestock. The **Southeast** includes four states, and it is the most developed region of Brazil. The population is distributed into very large metropolitan areas such as São Paulo and Rio de Janeiro, and mid-size cities. The economy is based on a strong and diverse industry, services, agriculture and livestock. Finally, the **South** includes three states, and its economy is based on automobile and textile, livestock and small

Table 1. The Distribution of programmers in the Brazilian geopolitical regions

Region	1st Quartile	Median	3rd Quartile	S. D.	Histogram
North					
Age (years)	20 to 30	20 to 30	30 to 40	0.72	
Education (degree)	Undergrad (ongoing)	Undergrad	Specialization	1.26	
Software Experience (years)	2 to 5	5 to 8	5 to 8	1.40	
OSS Experience (years)	1 to 2	2 to 5	2 to 5	1.03	
Northeast					
Age (years)	20 to 30	20 to 30	30 to 40	0.70	
Education (degree)	Undergrad (ongoing)	Undergrad	Specialization	1.23	
Software Experience (years)	2 to 5	2 to 5	5 to 8	1.05	
OSS Experience (years)	1 to 2	2 to 5	2 to 5	1.01	
Midwest					
Age (years)	20 to 30	20 to 30	30 to 40	0.61	
Education (degree)	Undergrad (ongoing)	Undergrad	Undergrad	1.04	
Software Experience (years)	2 to 5	5 to 8	8 to 12	1.26	
OSS Experience (years)	1 to 2	2 to 5	5 to 8	1.14	
Southeast					
Age (years)	20 to 30	20 to 30	30 to 40	0.65	
Education (degree)	Undergrad (ongoing)	Undergrad	Undergrad	1.16	
Software Experience (years)	2 to 5	5 to 8	8 to 12	1.24	
OSS Experience (years)	1 to 2	2 to 5	5 to 8	1.07	
South					
Age (years)	20 to 30	20 to 30	20 to 30	0.55	
Education (degree)	Undergrad (ongoing)	Undergrad	Undergrad	1.08	
Software Experience (years)	2 to 5	5 to 8	8 to 12	1.21	
OSS Experience (years)	1 to 2	2 to 5	5 to 8	1.13	

farm agriculture. Table 1 shows the user distribution per geopolitical region by age, education, professional software experience, and open-source experience.

We can see a number of interesting findings in the above table. First, as expected, we observed that the professional software experience is related to the OSS experience (the more professional experience, the more OSS experience). We attest this finding by running a Person Correlation ($r = 0.5654273$). Second, there is no correlation between age and the geopolitical region. Most of the respondents have between 20 to 30 years with little standard deviation in all regions. Third, as not expected, education degree is not related to open-source contribution ($r = 0.2257393$). Only the north and the northeast regions have more than 25% of their contributors holding a specialization degree. Nonetheless, the north group has only 15 samples (northeast has 137), and this sample size might represent a bias of the north population. Finally, we observed that most part of the contributions (62.38% of them) came from the southeast region. That is expected, due to its huge population size, as well as a whole range of universities and software companies located there.

3.2 Do the Brazilian Open-Source Contributions Increase over the Time?

We now examine the contributions on a monthly basis. During the analyzed period, we found more than 354,000 commits performed by 6,228 users in more than 98,000 projects. Figure 2 shows the number of commits per user per month.

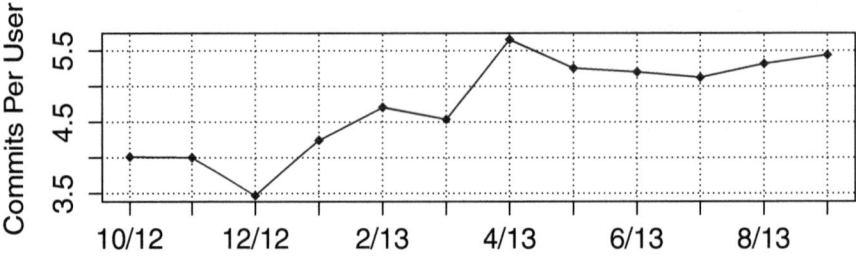

Fig. 2. The evolution of the absolute number of commits divided by the total of users

Firstly, the above picture shows that the contributions are increasing in absolute number – about 15.08% of increment during the analyzed period. One might think that the increment of contributions is because the active users are working more hard, and then they are performing more contributions. Nonetheless, we have observed that the peaks of contributions are due more users are willing to perform contributions in those months. Also, we have observed that the number of users that performs contributions in a month is about 25% of the overall active users. However, it does not mean that the same group of users will perform contributions every month. As we stated earlier, only 8% of the active users do contributions every month. Moreover, we observed that this number of users performing contributions per month can vary greatly. For instance, in August/2013, the month the received the highest number of contributors, the total of contributor is 25% bigger than December/2013, the month the received the lowest number of contributors.

Then, the absence of contributors is reason behind the low number of contributions in December and March. In December, besides be summer break in the universities, Brazil commemorates Christmas and New Year. So, usually, government and private industry provide one week off during this period. Moreover, the carnival, a big Brazilian holiday, occurred in March/2013. And again, during this period of time, universities and companies usually provide two, three days off. With this result we can assume that exists an open-source contribution trend in Brazil, and the contributors are willing to perform few contributions during their free time, but not on holidays.

3.3 Why do Brazilian Programmers Contribute to OSS?

Finally, as an attempt to understand why the Brazilian programmers do contributions to OSS, we analyzed the results of our last survey question: "Why do you contribute to OSS?". It was not a required question, but 81% of the respondents answered it. After the mining process, we found out more than 40 categories. Due to space constrains, we only describe most interesting ones.

The most common motivation factor was to **help to the community** with 19.40% of the answers. Some respondents said that *"To help people that have the same problem"*, and *"[My OSS contribution is] my 50 cents of contribution to the world"*, and *"I use a lot of OSS projects, so I do contributions to help someone else"*. We observed that the OSS spirit is very strong among respondents. Some of them believe that their contributions are not only about software, but a social contribution as well. This factor is related to the social aspects of the **Open source philosophy**, which has 10.88% of the answers. As an respondent described: *"Basically due to the OSS philosophy. I like the way of code sharing and I believe that it might improve my knowledge of [...]"*. On the other hand, some respondents do contributions because they want to **improve the software that they use** (17.75% of them). Most of these contributions are related to (i) fix simple problems or to (ii) implement new features. These findings are related to the work of Gousios [4], which found out that most of the contributions are consisted of a few lines of code. Then they advice contributors seeking to add a particular feature or fix to "keep it short".

Some respondents do contributions because they believe that they have to **give back the help received once** (17.63%). As a respondent said that *"Retribution, because I learned a lot from OSS communities [...]"*. Another motivation factor is to **improve their programming skills** (15.02%), which may happen during some code revision sections, and also improve their human capital by learning things besides programming. Then, the contributor might learn new concepts and good practices. However, despite the personal interest, we found out that only few respondents do contributions just to **improve their own curriculums** (6.27%) or to **gain visibility and reputation on the community** (4.49%). Due to the few number of respondents that have stated it as the main reason of their contributions, we assume that the majority part of the Brazilian OSS community are not motivated by self-marketing but by altruism and the fulfilment that arises from writing programs that other persons might use.

4 Related Work

There is a number of researches done regarding the Brazilian IT community, such as the Computer Science scientific production [1], the adoption of agile methods [11], the opportunities for women in IT jobs and education [10], among others. However, to the best of our knowledge, there is no study regarding the Brazilian open-source community in the literature.

Nonetheless, there are several studies regarding personal aspects of open-source communities. One study [5] investigated motivation aspects of 141 kernel developers. The authors revealed various motivational forces that contribute to a person's willingness to engage to OSS, both at the community level as well as the team level. In a similar study, Roberts et al. [13] develop a theoretical model relating the motivations, participation, and performance of OSS developers. They have reported a number of findings, including that developers' motivations are not independent but rather are related in complex ways. Also, they found out that different motivations have an impact on participation in different ways. In another study, Wang et al. [14] described a set of evolution metrics for evaluating open-source software (OSS) and community (OSC). They then measure the evolution of OSS and OSC together, and they showed that the Ubuntu success is due to the growth and maturation of its community.

Our work differs from the state of the art in two key ways. First, none of the above studies try to correlate their data using more than one data source. We argue that it is important because not always what the respondent say is true. We can then minimize this problem by double checking our findings in the two data sources. Second, to the best of our knowledge, it is the largest population size found in open-source studies. We believe that it is an important aspect, mainly because the use of large samples can increase precision and then reduce bias.

5 Conclusion

In this work we presented an empirical study concerning the Brazilian open-source contributor. We have analyzed an entire year of software development on Github, and we also conducted a survey with the active contributors. With the results of this mixed approach, we observed that: (i) the Brazilian open-source community is active. We have found that more than 54% of the population have performed at least one commit in a year. Nonetheless, as the Pareto's Law suggests, 20.35% of the users have performed 80.32% of the contributions; also, (ii) there is an open-source trend in Brazil. We noticed an increment of over 15% in the absolute number of contributions in one year, although about 24% of the active users perform commits monthly, and only 8% of them are active every month; and finally (iii) altruism is the main motivation, instead of self-marketing aspects. As a future work, we plan to understand the OSS adoption in other countries, and then, correlate our data with them. We also intent to collect more temporal data, as well as to gather information from other code hosting websites. Finally, we plan to investigate what are the reasons of open-source adoption in Brazil and South America as well.

References

1. Arruda, D., Bezerra, F., Neris, V., Toro, P., Wainera, J.: Brazilian computer science research: Gender and regional distributions. Scientometrics 79(3), 651–665 (2009)
2. Bagozzi, R., Dholakia, U.: Open source software user communities: A study of participation in linux user groups. Management Science 52(7) (2006)

3. Dabbish, L., Stuart, C., Tsay, J., Herbsleb, J.: Social coding in github: transparency and collaboration in an open software repository. In: Proceedings of the ACM 2012 Conference on Computer Supported Cooperative Work, ACM (2012)
4. Gousios, G., Pinzger, M., Deursen, A.: An exploration of the pull-based software development model. Submitted to the ICSE 2014 (September 2013)
5. Hertel, G., Niedner, S., Herrmann, S.: Motivation of software developers in open source projects: an internet-based survey of contributors to the linux kernel. Research Policy 32(7), 1159–1177 (2003)
6. Kitchenham, B., Pfleeger, S.: Personal opinion surveys. In: Guide to Advanced Empirical Software Engineering, pp. 63–92. Springer, Heidelberg (2008)
7. Kitchenham, B., Pfleeger, S., Pickard, L., Jones, P., Hoaglin, D., Emam, K., Rosenberg, J.: Preliminary guidelines for empirical research in software engineering. IEEE Trans. Softw. Eng. 28(8), 721–734 (2002)
8. Lin, L.: A Concordance Correlation Coefficient to Evaluate Reproducibility. Biometrics 45(1), 255–268 (1989)
9. McDonald, N., Goggins, S.: Performance and participation in open source software on github. In: CHI 2013 Extended Abstracts on Human Factors in Computing Systems, CHI EA 2013, pp. 139–144. ACM (2013)
10. Medeiros, C.: From subject of change to agent of change: Women and it in brazil. In: Proceedings of the International Symposium on Women and ICT: Creating Global Transformation, CWIT 2005, ACM (2005)
11. Melo, C., Santos, V., Katayama, E., Corbucci, H., Prikladnicki, R., Goldman, A., Kon, F.: The evolution of agile software development in brazil. Journal of the Brazilian Computer Society 19(4), 523–552 (2013)
12. Morrison, P., Murphy-Hill, E.: Is programming knowledge related to age? an exploration of stack overflow. In: Proceedings of the 10th Working Conference on Mining Software Repositories, IEEE Press (2013)
13. Roberts, J., Hann, I., Slaughter, S.: Understanding the motivations, participation, and performance of open source software developers: A longitudinal study of the apache projects. Manage. Sci. 52(7), 984–999 (2006)
14. Wang, Y., Guo, D., Shi, H.: Measuring the evolution of open source software systems with their communities. SIGSOFT Softw. Eng. Notes (November 2007)

Cuban GNU/Linux Nova Distribution for Server Computers

Eugenio Rosales Rosa, Juan Manuel Fuentes Rodríguez,
Abel Alfonso Fírvida Donéstevez, and Dairelys García Rivas

University of Informatics Sciences, School 1, Free Software Center
San Antonio de los Baños Highway,
Km 2 ½, Torrens, Boyeros, Havana, Cuba
{erosales,jfuentesr,aafirvida,dgrivas}@uci.cu
http://www.uci.cu

Abstract. This article presents the novelties offered by the new version of GNU / Linux Nova distribution in its server edition, exposing the new features such as network attached storage, distributed files system, charge balance for PostgreSQL database servers and thin clients, as well as the basic features of a standard server. All these developments are obtained from the integration with the server management platform Zentyal designed to facilitate the work of the end users of the variant of this Cuban distribution.

1 Introduction

After Informatics Havana 2009 event, Revolution Commandant Ramiro Valdés decided to implement an edition of the GNU/Linux Nova Cuban Distribution to migrate the country servers. A server oriented operating system must maximize the conditions of ICTs use in State Central Management Organisms (OACE) under technological sovereignty, security, socio-adaptability y sustainability principles [1]. For this reason, Nova development team releases that same year the first Cuban Server Distribution. This first version didn't count with a graphic interface to allow the administrator to easily and efficiently configure the system, so it was hard to manage the different services.

Besides, the fact that the administrator had to write directly in the configuration files might bring services errors, which often become fatal.

According to this situation, Nova team members wanted to facilitate servers management through a comfortable graphic interface, which could be also intuitive and highly reduce the occurrence of services errors on its next version.

Therefore, are proposed as objectives for this work:

- Integrate to Cuban Server Distribution a graphic interface for services management.
- Provide and develop a group of services as modules to the administration interface.

2 Development Methodology

OpenUP was used as software development methodology, because embraces a pragmatic and agile philosophy that centers on software development collaborative nature and allows more freedom because the model can be extended with part of other models, to face a wide variety of project types. It is a condensed process but is quite complete, with easy application to small and medium projects and easier to learn for smaller development teams. It is applied with an iterative and incremental focus inside a structured live cycle.

Management Platform: As a centralized management platform was used Zentyal, a unified open source network server created for small and medium enterprises (SMBs). It can act managing network infrastructure, as gateway to Internet, managing security threats, as office server, as unified communications server or a combination of them [2]. Besides, Zentyal includes a development framework to facilitate the development of new UNIX based services [3]. Even with all its options Zentyal does not have some services that are necessary in new informatics environments in the country, as SAN[1]/NAS[2] infrastructure and thin clients or no-disk machines. Fundamentally to these modules has been guided the work of the product developers.

Network Attached Storage Module: NAS is the name given to a storage technology dedicated to share the storage capacity of a computer (server) with personal computers or client servers through a network, using an optimized operating system to give access with *Microsoft Common Internet File System* (CIFS), NFS, FTP or TFTP protocols. NAS communications protocols are based in files, so the client requests the entire file to the server and handles it locally, that´s why they are oriented to information stored in small size and huge amount files. Using NAS provides some advantages as: capacity of sharing units, less cost, using the same network infrastructure with a simpler management. A plugin was implemented to Zentyal for adding support for implementing storage services on network via NAS, using ZFS[3], which is a file system developed by Sun Microsystems for their operating system Solaris. It has a 128 bits capacity (264 times a 64 bits file system capacity) [4]. For its implementation was necessary to add native support to Nova kernel.

Distributed Files System Module: Another product oriented to storage servers is the files systems cluster module, that allows to have multiple disks, even computers together in a cluster that is accessed through the network by client computers as a simple shared resource [5]. This method is even cheaper than NAS as thus can be implemented with outdated hardware already present in the country's institutions.

---

[1] Acronym of Storage Attached Network.
[2] Acronym of Network Attached Storage.
[3] Acronym of Zettabyte File System.

PostgreSQL Charge Balance Module: PostgreSQL is one of the free databases managers most used generally in the world and particularly in Cuba. It's common for a poor network infrastructure not to have powerful machines on which to mount this manager to be robust and reliable at all times. Nova Server provides an interface for PostgreSQL charge balancing that interacts directly with pgpool-II tool [6] and allows reducing the connection overcharge, and improve overall system performance; managing multiple PostgreSQL servers and reducing charge on them.

Thin Clients Suite Module: Nova has several ways of configuring thin clients, from the tedious manual configuration of all necessary services, to Osplugger: a native development that allows administration and configuration of most of this service related components. This tool, integrated with Zentyal, using LTSP module will allow Nova Server to gradually become the ideal suite for thin clients environments in the country, allowing Cuban operating system to be inserted gradually on network environments dominated nowadays by Windows Server or even other GNU / Linux distributions.

Antivirus Module: Zentyal by default uses ClamAV[4] antivirus, which is an anti-virus toolkit specifically designed to scan e-mail attachments in a MTA[5]. It has been said by MIC that "For virus protection, antivirus programs produced in the country will be used, or other officially approved for its use in the country, up to date." According to this, it has been developed a module for managing files and email through SavUnix[6].

Known Results: Once made and analyzed the proposed solution, GNU / Linux Nova Cuban Server Distribution was developed, and its integration with Zentyal, has made a GUI for telematic services management and to get a native product capable of deciding on it. It will be helpful for Cuban SMBs in the free software migration process that takes place in the country, as for an advanced alternative for network managers for convenient, simple and intuitive work. Being developed in Cuba contributes to its technological sovereignty, because it was possible to incorporate Nova Server's team as members and contributors of Zentyal development international community. It has also been established a direct and effective communication with the developers of the Cuban company Segurmatica Antivirus, managing to incorporate to the solution, a Cuban anti-virus, maintaining a constant and effective collaboration between both parties. This way not only has been possible to maintain a close relationship of collaborative development, but to give visibility to the project in the international framework. It has been possible to deploy thin clients on a smaller scale (10) in the University of Informatics Sciences (UCI) library and on a larger scale (64), in the MIC, working properly. Below is a table that reflect aspects that give validity to using Zentyal in Nova.

---

[4] Clam Antivirus: http://www.clamav.net/
[5] Acronym of Mail Transport Agent.
[6] Cuban antivirus solution.

Table 1. Tool comparison

	Windows	Webmin	YaST	Zentyal
Operating system	Windows	GNU/Linux	openSUSE	Nova
License	CAL	BSD	GPL	GPL
Usage easiness	High	Low	Medium	High
Executionmode	Local	Local	Local	Local
Security	Low	Medium	High	High

3 Conclusions

Nova Server is a project that has been gaining strength and is destined to be one of the products with more acceptance from Nova family, given its high levels of adaptation to the requirements of the Cuban enterprise informatization.

The development of new modules for its management system not only allows new features, but after reviewing the knowledge gained in its development, people can think of a sovereign software where decisional capacity and response times to errors are minimized, including compensation of this knowledge to the international community that can gain merits and recognition for Nova.

References

[1] Pierra Fuentes, A.: Distribución cubana de GNU/Linux: soberanía tecnológica, seguridad, criollo. Master Thesis, University of Information Sciences (UCI), Havana, Cuba (2011)
[2] eBox Technologies. Documentación de Zentyal 3.0 (2012), http://doc.zentyal.org/es/index.html
[3] Zentyal Module Development Tutorial (2012), http://trac.zentyal.org/wiki/Documentation/Community/Document/Development/Tutorial
[4] Zettabyte File System, ZFS (2012), http://ylvy.net/bloq/?p=467
[5] Montero, P., Misael, A.: Clúster de servidores web para aplicaciones desarrolladas sobre software libre que soportan altos niveles de concurrencia. Thesis project to qualify for the title of Engineer in Computer Science, University of Information Sciences (UCI), Havana, Cuba (2008)
[6] Sabater, J.: Replicación y alta disponibilidad de postgreSQL con pgpool-II (2008), http://linuxsilo.net/articles/postgresql-pgpool

A Study of the Effect on Business Growth by Utilization and Contribution of Open Source Software in Japanese IT Companies

Tetsuo Noda and Terutaka Tansho

Shimane University, Japan
nodat@soc.shimane-u.ac.jp, tansho@riko.shimane-u.ac.jp

Abstract. To analyze how OSS effects business growth both through simple use and by deeper engagement as a stakeholder in OSS community, we did questionnaire research to Japanese IT companies in 2012 and 2013. We analyze the progress of utilization and contribution of OSS, and the impact on business growth indicators by them.

1 Introduction

It has become commonplace for business enterprises to use OSS in their business activities. As a result, utilization of OSS only for cost reduction has become no longer a factor of obtaining the competitive advantages for them. The logic we understand as framing this such engagement is that the competitive edge that comes from technical advantages delivered by using OSS, and - using the same logic - it is therefore indispensable for them to contribute or participate in the development process of OSS. To verify this hypothesis, we questioned utilization and contribution of OSS, and business indicators to Japanese IT companies in 2012 and 2013.

2 Correlation between Utilization and Contribution of OSS

First, the correlations between utilization and contribution of companies in many OSS technology types are not shown in both years. As for these OSS technologies, Japanese IT companies are still "free riders." On another front, the correlation between Ruby and Ruby on Rails in this context is shown. And those of Other Languages between Apache and Databases technologies are also shown.

Table 1. Correlations between utilization and contribution of OSS

contribution / utilization	Linux	Apache	Databases	Ruby	O.L.	RoR	contribution / utilization	Linux	Apache	Databases	Ruby	O.L.	RoR
Linux	.136	-.002	.004	.128	.083	.110	Linux	.062	.009	-.013	.080	.032	.037
Apache	.151	.135	.054	.149	.125	.111	Apache	.072	.048	.012	.095	.013	.033
Databases	.050	-.016	.052	.132	.098	.105	Databases	.136	.086	.106	.083	.082	.043
Ruby	.031	-.013	.007	.324**	.114	.351**	Ruby	.098	.057	.044	.461**	.164	.320**
Other Languages	.144	.161*	.189*	.099	.272**	.140	Other Languages	.139	.114	.144	.208*	.217*	.257**
Ruby on Rails	.087	.086	.065	.331**	.159	.420**	Ruby on Rails	.180*	.141	.129	.400**	.239*	.343**

Survey Results in 2012 (n=191)   Survey Results in 2013 (n=146)

Spearman's rank correlation coefficient   ** 1% level of significance, * 5% level of significance

3 Effect on Business Growth by OSS Utilization and Contribution

From the survey result in 2012, the subsequent period prospect of sales growth rate is impacted by utilization of OSS. At the same time, the survey result in 2013 shows that the present period of sales growth rate is impacted. It turns out that the sales growth rate prediction by utilization of OSS expected from the reference fiscal year appears in the following fiscal year.

Table 2. Correlations between business growth and utilization of OSS

	Growth Rate of Sales (present period)	Prospect of Sales Growth Rate (subsequent period)	Growth of Employee Number (present period)	Prospect of Employee Number's Growth Rate (subsequent period)
Linux	.191**	.245**	.207**	.133
Apache	.167*	.220**	.079	.066
Databases	.131	.222**	.026	.067
Ruby	.135	.214**	.063	.113
Other Languages	.098	.176*	.052	.092
Ruby on Rails	.055	.178*	.061	.068

Survey Results in 2012 (n=191)

	Growth Rate of Sales (present period)	Prospect of Sales Growth Rate (subsequent period)	Growth of Employee Number (present period)	Prospect of Employee Number's Growth Rate (subsequent period)
Linux	.302**	.194*	.159	.091
Apache	.189*	.113	.129	.071
Databases	.306**	.219*	.201*	.134
Ruby	.207*	.148	.149	.106
Other Languages	.237**	.125	.164	.053
Ruby on Rails	.171*	.098	.132	.044

Survey Results in 2013 (n=146)

Spearman's rank correlation coefficient ** 1% level of significance, * 5% level of significance

Table 3. Correlations between business growth and contribution of OSS

	Growth Rate of Sales (present period)	Prospect of Sales Growth Rate (subsequent period)	Growth of Employee Number (present period)	Prospect of Employee Number's Growth Rate (subsequent period)
Linux	-.091	.007	-.032	-.089
Apache	-.031	.021	-.092	-.127
Databases	-.036	.092	-.083	.020
Ruby	.052	.047	.072	.058
Other Languages	.019	.057	-.029	.002
Ruby on Rails	.034	.075	.018	.049

Survey Results in 2012 (n=191)

	Growth Rate of Sales (present period)	Prospect of Sales Growth Rate (subsequent period)	Growth of Employee Number (present period)	Prospect of Employee Number's Growth Rate (subsequent period)
Linux	.001	-.003	.108	.219*
Apache	.023	.018	.054	.215*
Databases	-.029	-.018	.071	.182*
Ruby	.000	-.053	.054	.115
Other Languages	.051	-.025	.127	.206*
Ruby on Rails	.004	-.039	.046	.135

Survey Results in 2013 (n=146)

Spearman's rank correlation coefficient ** 1% level of significance, * 5% level of significance

The contribution of OSS communities has an insignificant effect on sales growth. But he survey result in 2013 shows that the prospect of employee number's growth rate (subsequent period) is impacted by contribution of OSS. Japanese IT companies tend to expect direct sales growth by OSS practical use, and they might also expect contribution to the development process of OSS leads to the increase of employee, personnel training, and personnel adoption.

Reference

1. Noda, T., Tansho, T., Coughlan, S.: Effect on business growth by utilization and contribution of open source software in japanese IT companies. In: Petrinja, E., Succi, G., El Ioini, N., Sillitti, A. (eds.) OSS 2013. IFIP Advances in Information and Communication Technology, vol. 404, pp. 222–231. Springer, Heidelberg (2013)

USB Device Management in GNU/Linux Systems

Edilberto Blez Deroncelé, Allan Pierra Fuentes,
Dayana Caridad Tejera Hernández, Haniel Cáceres Navarro,
Abel Alfonso Fírvida Donestévez, and Michel Evaristo Febles Parker

University of Informatic Science, Free Software Center,
Road to San Antonio de los Baños, 2 ½ Km, Torrens, La Havana, Cuba
{eblez,apierra,dtejera,hcaceres,aafirvida,mfparker}@uci.cu
http://www.uci.cu

Abstract. Protecting the access to USB ports has the same priority for information security than firewalls and antivirus software. Nowadays there are some tools that allow us to monitor and regulate the access to USB devices, but all of them are distributed under proprietary licenses. This work presents an application that solves the mentioned problem: ¿How controlling the access to USB mass storage devices in GNU/Linux Operating Systems?

1 Introduction

One of the most important advantages of devices guided by USB industrial standard is the access easiness to every kind of computer that provides its usage. Theoretically this can be an advantage to the enterprises, except for the fact that concepts "security" and "access" are completely opposite in the Security Information area.

After the creation of the Universal Serial Bus, at the beginning of 90s, the development of technologies around of USB devices has been constantly evolving. The most recently versions of this kind of devices have increase its storage capacity, which implies the improvement of its performance and the decrease of its height.

So, with USB devices user can count with an easy means of transport, to store information while maintaining its availability. Anyway, the easy access of these devices to computers and the sensitivity of the information handled by them, can bring a set of disadvantages and vulnerabilities, both for end users and businesses. Deliberate or accidental users can:

- Remove critical data.
- Exhibit confidential information.
- Introduce malicious code that can affect the entire corporate network.
- Transferring inappropriate or offensive hardware company material.
- Make personal copies of company information and intellectual property.
- Connect portable devices and consequently distract workers during working hours.

In an attempt to control these threats, companies began to prohibit use in workstations of USB devices for personal use. However, practice indicates that trusting on voluntary compliance by users is not enough. The best way to monitor and regulate access, and maintain control of these devices has been by placing technological barriers.

Migration to open source technologies and open standards in the companies has found barriers in this regard. The protection of information and the USB ports of computers was made in companies using proprietary applications like "MyUSBOnly", "GFI EndPointSecurity" and others. How it has not been found a similar application in the field of free software. This research persues the goal of present a tool similar behavior but to enviroments GNU/Linux. The version we have today in GNU/Linux specifically Cuban GNU/Linux is the Unix Smart Blocker for Universal Serial Bus, USB^2 1.0 tool, which is the application shown in this paper.

2 Contents

A technological barrier to the uncontrolled use of USB devices in a company is the solution proposed in this research. The application USB^2 1.0 will allow controlling and restricting the use of USB devices on computers that use GNU operating systems.

The code of the application currently in development can be located https://github.com/editoblezd/usbsecurity-dev

2.1 Methodology

During this research were used scientific methods: analysis-synthesis, historical analysis and experimental one, which allow us the foundations for the development of the application.

By Analysis-synthesis method were studied the most used technologies nowadays to identify its main features; which were the basis for the design and implementation of the first version of USB^2 1.0. Some of them were: control of USB ports via a database with authorized devices and alerting users to the authorization or not of new devices.

Historical analysis method allows us to study the evolution of this kind of application and as a result, to identify the need that existed in the distributions of GNU/Linux for a tool to control access to USB storage devices. University of Informatics Sciences (UCI) users and the free software community were the main samples for this research.

The experimental method allowed testing the correction of the functionalities implemented with main samples.

The development of the application was useful to the user community GNU/Linux, because is an alternative to protect users from data loss, data robbery, or other negative consequences of the use of USB mass storage devices. Once developed the tool, safety levels in environments GNU increased and hence the users' trust in free operating systems.

2.2 Results and Discussion

Some of the tools used to control USB devices are: USB Blocker, Device Lock, GFI EndPoint, MyUSBOnly, Endpoint Protector 4 and USB Over Network Server; but most used are the first five.

The freeware software USB Blocker created by NetWrix Corporation, is an interesting centralized locker that allows controlling access to the computers on a network using enabled USB devices. It locks different kinds of mass storage devices, such as removable hard disks, iPods and others [1].

USB Blocker can control the access of authorized or unauthorized devices, functioning as a mean of protecting security control malware, viruses or loss of sensitive information for the corporate network where is used. Unlike other solutions, does not require installation on each computer on the network, functioning as an integral network control system. Its main disadvantage is that in addition to be proprietary is conceived only for Windows systems.

Device Lock allows defining which users can access to specific ports and devices on a computer. Its management can be made using group policies in an Active Directory domain, creating a console for administration [2]. But, like USB Blocker, is available only for Windows platforms NT/2000/XP/Vista/7 and Windows Server 2003/ 2008, and is proprietary. Among the main features that can be managed with Device Lock are: users or groups access control, a whitelist of devices regardless of where they connect, ports auditing and integration of TrueCrypt and PGP Whole Disk encryption.

GFI EndPoint is another powerful tool that has among its main functions: group-based protection control, granular access control , support for various kinds of portable devices, logging user activity in SQL Server , the agent protection with increased security, control of portable storage with Unicode support and the event log [3].

MyUSBOnly provides the least amount of functionalities, but it is the most used by companies and end users. Allows quick and easy way to protect USB ports and set a password to access them. Requires Microsoft Windows 2000, XP, 2003, Vista, 7 or 2008 [4] as operating system.

All these applications are developed for Microsoft Windows operating systems which makes slower the migration process of companies to free software.

Currently, the tool Endpoint Protector 4, brand development some guidelines for tool described in this paper. Although this tool solves the problem posed in this investigation, but it is not distributed under GPL license, which becomes a obstacle for GNU environments.

The Endpoint Protector 4 Web administration and reporting console offers a complete overview of the device activity on your computers, whether you work with Windows, Mac or Linux platforms. The enterprise will be able to define access policies per user/computer/device and authorize devices for certain user or user groups. Thus, the company will stay productive while maintaining control over the device fleet use.[5]

The tool presented in this paper aims to be fast, efficient and simple. It has good acceptance by Cuban users and is compatible with GNU/Linux systems and is distributed under GPLv3.0 License. The application libraries are distributed under the LGPL v3.0 license. USB[2] 1.0 increases safety levels required in Cuba and its usage can be included in the safety regulations must have any cuban institution under Resolution No.127/2007 [6].

2.3 Development Technology

For the development of the application was used C, as the programming language, using glibc libraries and gtk +-2.0 for interfaces. Was used vim as Integrated Development Environment and to store information in the database is used squlite3 library.

USB[2] 1.0 1.0 has incorporated a default graphical user interface in text mode. The tool, developed as a library, provides the necessary tools so you can build a graphical interface for different environments.

2.4 Recognition of the Device in the Computer

Udev is the device manager for the Linux kernel. Essentially this device handles devices nodes that are regularly can be found in /dev directory. This is the successor of devfs and hotplug, so it handles /dev directory and all actions taken in user space when devices are added and subtracted in the computer.

How udev by default is active for all systems that use the version 2.6.x of the kernel or upper, is used to recognize devices in the system. In the process of connecting or disconnecting a device to the system, udevd demon executed by udev receives a uevent released by kernel indicating that an action of adding or disconnecting a device has been captured.

Sysfs is a virtual file system that provides the Linux kernel v2.6. Exports information about devices and drivers from the kernel device model to userspace and also allows to set its parameters. When a device is attached to the system, in addition to the information stored by udev in its database, the kernel enclosed information necessary for its work in the virtual/sys directory.

Among the possibilities of udev are its rules. These rules have a syntax that allows both read attributes of a device and change others, depending on the type of attribute. A set of rules decide what action needs to be done with obtained data. What rule will do, probably, will be to name the device, to create the appropriate device file, and run the program that has been set to activate the device.

Rules can associate a unique name to a device, but can also call an external program to give more information about the device; so that can obtained a more specific name. Most of the rules that come by default in the system are stored in /lib/udev/rules.d/.

By using these rules is performed a device recognition and are selected only storage devices to manage access to the system. The attributes that are granted to allow uniquely identify devices and thus can control them.

The device recognition is an important phase in its access control process system to the system. If this action is performed successfully can be achieved the authorization of only those devices that have been previously authorized by the administrator or superuser.

The rule would look like this:

> #Regla para usbsecurity1.0
> ACTION=="remove",GOTO="disk_usb_end"
> SUBSYSTEM!="block", GOTO="disk_usb_end"
> # Para importar información de los padres.
> ENV{DEVTYPE}=="partition",IMPORT{parent}="ID_*"
> #Ignorar discos floppy
> KERNEL=="fd*",ENV{UDISKS_PRESENTATION_HIDE}="1"
> KERNEL=="sd*[0-9]|sr*",ENV{ID_SERIAL}!="?*",SUBSYSTEMS=="usb",IMPORT{program}="usb_id --export %p"
> #Ejecutar un programa al conectar dispositivos USB.
> KERNEL=="sd?",ACTION=="add",ENV{ID_BUS}=="usb",SUBSYSTEM=="block",SUBSYSTEMS=="usb",PROGRAM!="/usr/sbin/usbsecurity /dev/%k", ENV{UDISKS_PRESENTATION_HIDE}="1"
> ENV{DEVTYPE}=="disk",   KERNEL!="sd*/sr*",
> ATTR{removable}=="1", GOTO="disk_usb_end"
> LABEL="disk_usb_end"

As can be seen when a storage device is connected, this is the action is ACTION = "add", the "usbclient" program is executed passing the name of the device node recognized by the kernel as argument. This program performs device access check tasks and returns true or false depending on whether the device is not authorized to access the computer or if it is respectively. In case of been authorized, the rule does not reach its end and the device goes through the normal connection and removes process. Otherwise does not allow accessing the computer by removing the device node so that can not do access device to the system even with the root password.

2.5   Verification of Access Permission

For verification of system access, is necessary to know several device attributes such as serial number. For getting such attributes are used programs already offered by udev as usb_id which exports device information to the udev environment variables, in this case specifically are exported the attributes with udev format and the order:

> IMPORT{program}="usb_id --export %p", escrita en la regla de udev si es que otro programa no la han importado ya. La sección que comprende esta orden ejecuta de forma general:
> # USB devices use their own serial number
> KERNEL=="sd*[!0-9]/sr*",
> ENV{ID_SERIAL}!="?*",SUBSYSTEMS=="usb",
> IMPORT{program}="usb_id --export %p"

Once imported to this information can access this in the written program usbsecurity.c using the function: g_getenv(ENV_SERIAL), which returns the serial number of the device. In addition is needed to know what type of device it is, this is if it is a partition or a disk. That is obtained through the g_getenv(ENV_DEVTYPE) function which reads this environment variable.

Once imported this information, application checks if the serial device is already registered in its database, thus deciding the return value. This return value depends on whether the device is finally shown to the user or not.

2.6 Managing Devices

The application has a graphical user interface with which the devices are managed, a new device can be added to the database, can be deleted and modified its identifier too. By default when a device is connected and it is not authorized is recorded in the database but is not authorized. In the ID field is set by default Not allowed, which means that although this is not registered can not access the system until you can not change the ID.

Adding a Device

To add a device must be connected to the computer and if it is not authorized, will show the user a notification to let him know the current state of the device. To authorize a system, administrator must access the management interface for the device approval process.

In this interface the administrator must change the default identifier that takes the device, to authorize it. If, however, you will not change this ID after log all devices that still have the original identifier will be deleted from the database. Once the device is authorized it will be mounted automatically in the file system of the operating system. Once the device is authorized, is launched to the user a notification informing the changes.

Sorting an Authorized Device Identifier

The process of modifying the device ID, performs in the same way that of adding a device to the system. The difference lies in that the device is already authorized in the system and as a first step the administrator or super user had to change the default identifier.

Removing a Device

To remove the device, simply must be selected in the administration interface and click the button to the elimination. After acceptance of the confirmation message, the device will be removed from the database and will automatically unauthorized use in the system.

Activity Logging

Internally the system, in addition to the information displayed in the Admin GUI, store other attributes in system logs. When any activity, both connection authorization and de-authorization is performed, is stored information relating to this as time, date, and user who performed the action and the serial number of the device. This helps finding events if an abnormality arises, to find the root of the problem. It is important to note that how this valuable information is stored, it can be known information about any device that attempts performing any action, which ensures an extra level of security, in this case the detection.

3 Conclusions

USB Unix Smart Blocker 1.0 provides another point of support for the growing process of migration to open source platforms. With this version of the product the administrators of workstations will be able to control the access of USB devices, increasing levels of security for enterprises.

From the economic point of view, for a country like Cuba, paying for software licenses over 600,000 workstations migration process is a rather high figure. USB Unix Smart Blocker 1.0 is distributed under GPL license, besides being a value-added product to the Cuban distribution of GNU / Linux Nova, used as the main distribution in the migration process of the country.

The architecture used in the development of the application allows for future expansion and scalability of it. What provides companies with a tool adaptable to the conditions they need.

USB^2 1.0 has arrived as a free alternative in GNU/Linux systems to monitor and control USB devices.

4 Recommendations

It is recommended:

1. To expand the number of devices that are supported by this tool, to have control of these.
2. To use this tool to protect your USB ports, measure as important as installing a firewall and antivirus program maintenance. USB^2 1.0 makes easy the protection of this critical part of your PC or laptop.

References

1. Alejandro, J.: USB Blocker, eficiente bloqueador de acceso USB para red (2009), http://www.acercadeinternet.com/usb-blocker-eficiente-bloqueador-de-acceso-usb-para-red/ (accessed: October 29th, 2012)
2. PYME. DeviceLock, control total sobre los USB en la empresa (2010), http://www.tecnologiapyme.com/software/devicelock-control-total-sobre-los-usb-en-la-empresa (accessed: October 29th, 2012)
3. GFI Endpointsecurity. Feature (2006), http://www.gfi.com/usb-device-control#features (accessed: October 29th, 2012)

4. MYUSBONLY. System Requirements (2012), `http://www.myusbonly.com/usb-security-control-device-protection/` (accessed: October 29th, 2012)
5. Endpoint Protecor 4 (2013), `http://www.endpointprotector.com/products/endpoint_protector#1` (accessed: December 26th, 2013)
6. Ministerio De La InformáTica y Las Comunicaciones. Resolucion No. 127 (2007) (accessed: October 29th, 2012)

PROINFODATA: Monitoring a Large Park of Computational Laboratories

Cleide L.B. Possamai, Diego Pasqualin, Daniel Weingaertner, Eduardo Todt,
Marcos A. Castilho, Luis C.E. de Bona, and Eduardo Cunha de Almeida

Centro de Computação Científica e Software Livre,
Departamento de Informática - Universidade Federal do Paraná
{cleide,dpasqualin,danielw,todt,marcos,bona,eduardo}@inf.ufpr.br

Abstract. This paper briefly presents a model for monitoring a large, heterogeneous and geographically scattered computer park. The data collection is performed by a software agent. The collected data are sent to the central server over the Internet, and stored by the storage system. An on-line portal makes up the visualization system, featuring charts, reports, and other tools for assessing the state of the park. This system is currently monitoring circa 150,000 machines.

1 Introduction

Digital inclusion is a major challenge of modern society. The Brazilian government has made large investments in equipment and infrastructure to grant access to information and communication technologies. With the success of installing laboratories in a large territory, arises another challenge: to monitor such giant park, ensuring that all investment meets its social and economic objectives. Aiming to solve this problem, it is necessary to define a model and implementation of IT solutions that enable the monitoring, in an easy and transparent way.

2 System Description

The proposed system has three independent modules: *i*) Data Collection System; *ii*) Data Storage System; and *iii*) Visualization System.

2.1 Data Collection

The collection module gathers data from computers spread over the country. A client-server scheme is adopted. The agent implements the client Web Service, collecting the information from the source machine and storing it locally in XML format, connecting to the central server and sending the data. The central server collects this data from all agents, storing them in a database and also providing authentication methods and secure connections.

The implementation of the Web Service client was done in C due to performance issues and minimization of the number of packages that must be sent, since in many regions of Brazil the quality of internet connection is poor. The Web Service server is an Apache Axis because it contains the base protocols and safety extensions implemented. Due to the large amount of machines, the server does not maintain the connection states of the clients.

The computers are distributed by the ministry of education with Linux Educational (http://linuxeducacional.c3sl.ufpr.br/), with the monitoring agent installed. Since schools may install another operating system or distribution, the agent has been designed to work across major operating systems and distributions.

In Linux-based systems, data collection was developed using Shell Script language, whereas on machines with Windows operating system, data collection was developed using the Python programming language. In order prevent the server from being overloaded, each agent sends the inventory at random times, distributing the total load along all the day. The data collected by the agent are the identifier code of schools, type of machine, network usage and inventory.

2.2 Storage System

This module uses a typical storage architecture for large-scale data (Data Warehouse - DW), oriented to read operations that favors the analysis of large volumes of data. A DW architecture consists of three stages: loading, storing and reading data. This division increases the overall performance of the system where different operations (i.e. writing, reading) are directed to different structures. This also adds versatility, so that it is possible to replace one technology by another without impact on the rest of the system, e.g., through the use of new technologies that may further improve data read performance, e.g., MapReduce [3], and C-Store [4].

The three stages are divided in two separate databases: a temporary data database (Staging Area) and a database of historical records (Data Warehouse).

The Staging Area is a database that receives connections from the web server to write data collected during the day. This database has its data deleted after the consolidation of daily data held by DW. The Data Warehouse is a database responsible for storing all historical data. DW reads and consolidates the data from the staging area, once a day, starting at predefined times.

The Data Mart is a subset of DW, summarized and optimized for reading operations [1]. These subsets are determined according to the interest of the consultations. The goal of DMs is to increase the overall performance by creating specific structures for each universe of reports, such as inventory, availability, audit, and network usage. The database was implemented using technology based on open source, in our case, with the management database system (RDBMS) PostgreSQL (postgresql.org). What makes Data Warehouse a high performance system is that the information is joined in a database in a dimensional way [2], reducing the use of join operations in queries.

2.3 Visualization System

The visualization system is composed of a web portal, which is in turn subdivided into a server and a client. The web server is responsible for receiving and processing requests, fetching the requested data from the database (Data Marts), and returning the report or graph to the client. The client application is the web portal itself, viewed through a browser and responsible for presenting an interface that facilitates the search and visualization of monitored data.

The proposed model is based on a multilevel grouping of monitored machines, for example, arranged by geographic location and institution. The visualization system has in the first level an overview of the machine park, showing the total number of monitored machines, or any other aggregation meaningful to the administrator. A second level provides information classified by federation States. Selecting in a third level, the user can view data from a specific city and at the last level the machines are displayed individually.

The first version was developed with the Pentaho (www.pentaho.com) tool, but due to the lack of some desirable native features, like caching, resulting in a noticeable slowness in requisitions answers, it was replaced by the Play Framework (www.playframework.com) with the Highcharts (www.highcharts.com).

3 Experimental Results

All the tools used in the project are open source software, as well as the final product delivered (http://git.c3sl.ufpr.br). The result of the implementation can be seen in the portal http://proinfodata.c3sl.ufpr.br.

Figure 1 presents the visualization of availability information of the computer park for a region (a) and for a federation State (b). The pie chart shows global information, whilst the bar chart groups the data by region, and the line chart shows the history of the last six months. When selecting a city, a report containing the data from all machines of monitored schools within that city is generated.

Currently, the number of monitored computers with Linux Educacional is 139,045, with Windows is 246, and with other systems is 4,323, summing up to 143,614 machines. The staging/dw model was tested with approximately 200,000 connections per day, writing data individually and in parallel.

4 Discussion and Conclusions

The results show that the proposed model and the subsequent implementation are effective and allow the administrator to manage the set of laboratories distributed over the country, based on a system that is optimized to support the giant number of clients that should send status information to a central server. The hierarchical approach makes easy to the user to locate the information needed. The color codes used to identify the groups of machines according to the time elapsed since the last contact make easy to recognize problems such

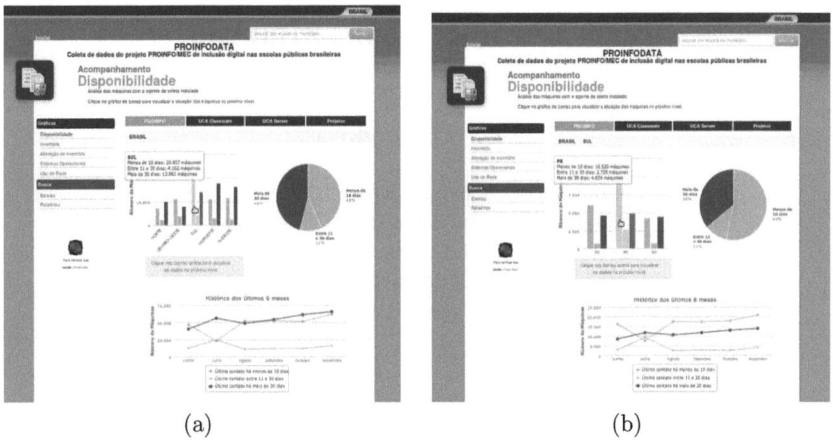

Fig. 1. Data about availability grouped by Region (a) and State (b)

failures and machines that were not yet put into operation. According to the Data Mart model, the information of individual machines is consolidated calculating *a priori*, once a day, subtotals for cities, States and regions, accelerating the responsiveness of the system to user queries. One fundamental characteristic is that the system is open source, allowing code inspection and assurance of privacy about personal data. As future work, it could be added artificial intelligence modules able to find anomalous behaviours and to generate alarms to the administrator. Also, since the system is modular and designed to be platform independent there are few obstacles to continuous improvements.

References

1. Designing data marts for data warehouses. ACM Trans. Softw. Eng. Methodol. 10(4), 452–483 (2001)
2. Bellatreche, L., Woameno, K.Y.: Dimension table driven approach to referential partition relational data warehouses. In: Proceedings of the ACM Twelfth International Workshop on Data Warehousing and OLAP, DOLAP 2009, pp. 9–16. ACM, New York (2009)
3. Lämmel, R.: Google's mapreduce programming model – revisited. Science of Computer Programming 70(1), 1–30 (2008)
4. Stonebraker, M., Abadi, D.J., Batkin, A., Chen, X., Cherniack, M., Ferreira, M., Lau, E., Lin, A., Madden, S., O'Neil, E., O'Neil, P., Rasin, A., Tran, N., Zdonik, S.: C-store: A column-oriented dbms. In: Proceedings of the 31st International Conference on Very Large Data Bases, VLDB 2005, pp. 553–564. VLDB Endowment (2005)

Book Locator: Books Manager

Dairelys García Rivas

University of Informatics Sciences, School 1, Free Solutions Center,
San Antonio de los Baños Highway, Km 1 ½,
Torrens, Boyeros, La Habana, Cuba. CP: 19370
dgrivas@uci.cu
http://www.uci.cu/

Abstract. After performing a study of digital books organizing tools, it's observed that these don't count on a multi platform integration, and those who do, don't count on the elements pursued in the investigation. It is decided then to proceed to the implementation of a book organizer software, taking an initial requirements list from the studied tools, following the guidelines of the collaborative and open source development. It has been developed to its version 4.1, which after going through different development processes, detecting mistakes and adding new functionalities that join to the requirements list, it's fully functional.

1 Introduction

With the increasing use of information technologies and Internet, it becomes increasingly common to use digital documents to save the information. Books are not an exception; since the growth of these technologies, people have tended to favor digital reading. This is also linked to the development of mobile devices, which allows transporting digital books. Because of this, today is much more convenient to use an Ebook Reader[1], since they can carry thousands of books simultaneously in a confined space; plus the cost of digital books is much lower than physical books.

For this situation, there have been programs that allow reading these documents, in different platforms and to various formats, being most common the PDF format. These programs, however, have characteristics that prevent people working with them on multiple operating systems. Taking into account the characteristics of these applications, summarized in Table 1: Comparison of locating books programs, will be possible to decide the most appropriate program for working on multiple platforms.

1.1 Known Results

The intended result is based on a layered architecture, in this case two layers. It uses a presentation layer to handle all the elements involved in the generation of system views (Add, Delete, Advanced) and the warning and error messages. The other layer

---

[1] A term that refers both to an electronic device used to read books in digital format.

Table 1. Comparing books locating programs

Name	Language	Platform	Special features	Supported formats
mCatalog	English	Windows	-	All
Alexandria	English	Linux	Used mainly in Ubuntu, gets broken sometimes	All
Calibre	Spanish	Linux, Windows, OS X	Disorganized graphic interface	All
FBReader	Spanish, English	Linux, Windows	Mainly designed for mobile devices	Epub, html, mobi, palmDoc, zTxt
Tellico	Spanish, English	Linux (KDE)	Can't work properly in Ubuntu and with GTK interfaces	All

used focuses on controlling access to data, validating the data before inserting it into the database and also performs queries to it, returning the results to the presentation layer.

In the main menu, there are available the options "Add book", "Delete book", "Edit book", "Advanced Search" and "Exit". The latter option terminates the application execution. In the "DB" menu are the database management options (" Create new BD", "Insert / Remove Language", "Insert / Remove Format" , "Insert / Remove genre") and in the "Utilities" menu are other utilities that the user might need ("Amount of read books", "To read books list", "Genre / Format / Language most read" and "Credits").

The interface also has a tab panel, where you can find all the books inserted, sorted in alphabetical order or by the author who wrote it. When double-clicked on the author's name, the names of the books he has written will be displayed. At the top of the window you there is a search by book name or by the name of the author who wrote it; in each case the results are shown in the panel tab. On the right side you can see the book information once you have selected in the left tab. In the bottom of the window there is a text element to be updated with interesting facts the author, as the total number of books out there, or how many books he has written an author if you select it on your tab.

The program has been tested by a group of developers of the Operating Systems Department in Cesol Center, Faculty 1. They have positive opinions regarding its functionality and support multiple platforms. The views on this have varied depending on the personal tastes of the users and the features that they want to have in a program. These recommendations have been reported to be taken into account for future releases, and it has been proposed to be inserted in the repository of the Cuban GNU / Linux distribution, Nova.

Automation of Agricultural Irrigation System with Open Source

Bladimir Jaime Pérez Quezada and Javier Fernández

Faculty of Computer Science, Electronics and Communication,
Universidad de Panamá, Panamá
bladimir.perezq@gmail.com,
fernanj@yahoo.com

Abstract. In this present job, we seek to develop a prototype of an automated agricultural irrigation system, monitored and controlled remotely. For that, we will use inexpensive tools, flexibility and support such as Arduino, XBee and Android. Arduino and XBee, will be responsible of the automating the system. Android will achieve the remote monitoring and control from anywhere in the world where there is cellular service and Internet. In this way we can give the farmer the comfort and security that he don´t get with a manually controlled system.

Keywords: Arduino, ADK, Android, XBee.

1 Introduction

At present irrigation systems, the vast majority are operated manually, requiring close attention of the farmer. Basically the farmer decides with your experience, the period and the moment to watering the crop. Because of this, will can have common problems caused by humans, such as forget turn on and turn off the system at the time considered. This causes direct problems with the crop such as, excess of water in the crop or drying by lack thereof, in addition to a higher energy consumption of the system and higher consumption of water resources that is as important to national and international level. All this lowers productivity, causing economic losses to farmers. Having an automated system, we seek to resolve these errors, increasing production, save on energy consumption and minimize the decline in productivity, as the farmer will not has that stay aware constantly of the cultivation, because the system will be Stringer and it will take the required decisions [2].

2 Methods and Materials

To develop our agricultural irrigation we focus on aspects such as: open source, scalability, open to future research and implementations. The tools used in our research are:

- **Arduino Mega ADK.** Integrates one USB host that work directly with Android devices and it will be responsible of take the decisions pertinent of the irrigation system.
- **Android.** Will show the details of the system to the user and serve as link to the Internet to upload data to the cloud system of Google Drive.
- **XBee.** Modem based on ZigBee wireless communication protocol. Developed and backed by the company Digi International. Will maintain communication with the Arduino Mega ADK, sending the soil moisture data.
- **Soil Moisture Sensor.** Soil moisture sensor of resistive type, whose operating principle is based on the conductivity, which will vary depending on the moisture present in the soil [1].

3 Expected Results

Get an agricultural irrigation system low cost, that can be implemented in real crops, offering the water required by cultivation. That the user has the same monitoring both on the site as remotely via the Internet through a computer or mobile device. Furthermore, the user can control the system remotely via SMS, as shown in Figure 1.

With this research, we want to help the farmer get a better production and minimize waste.

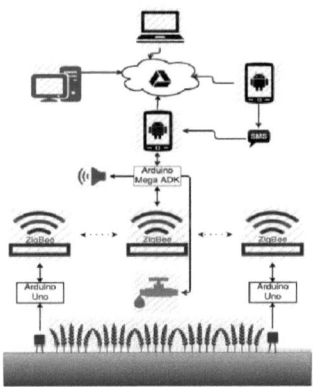

Fig. 1. Functional diagram of the system

References

1. Schugurensky, C., Capraro, F.: Control Automático de Riego Agrícola con Sensores Capacitivos de Humedad de Suelo, aplicaciones en Vid y Olivo, Universidad Nacional de San Juan, Argentina. Instituto de Automática, INAUT (2007)
2. Varela, R.V., Rivas, C.R., Aréchiga, R.S.: Automatización de un sistema de riego. Revista Digital de la Universidad Autónoma de Zacatecas, Universidad Autónoma de Zacatecas, Enero, Abril. Unidad Académica de Ingeniería Eléctrica (2007)

When Are OSS Developers More Likely to Introduce Vulnerable Code Changes? A Case Study

Amiangshu Bosu[1], Jeffrey C. Carver[1], Munawar Hafiz[2], Patrick Hilley[3], and Derek Janni[4]

[1] University of Alabama
[2] Auburn University
[3] Providence College
[4] Lewis & Clark College

asbosu@ua.edu, carver@cs.ua.edu, munawar@auburn.edu, philley@friars.providence.edu, derekjanni@lclark.edu

Abstract. We analyzed peer code review data of the Android Open Source Project (AOSP) to understand whether code changes that introduce security vulnerabilities, referred to as vulnerable code changes (VCC), occur at certain intervals. Using a systematic manual analysis process, we identified 60 VCCs. Our results suggest that AOSP developers were more likely to write VCCs prior to AOSP releases, while during the post-release period they wrote fewer VCCs.

Keywords: Open Source, OSS, FOSS, security, vulnerability.

1 Motivation

The presence of a security vulnerability in a widely used software, like Android, can be critical. Therefore, project management will be able to make informed decisions on allocating scarce security experts, if they can predict 'where', and 'when' vulnerable code changes (VCC) are more likely to be introduced. Most of the prior studies [1,2] on vulnerability prediction considered the 'where' aspects. To investigate the 'when' aspects, this study analyzes 60 VCCs from the Android Open Source Project (AOSP) to determine when AOSP developers are more likely to introduce VCCs.

2 Study Design

The AOSP project uses the Gerrit code review system for changes submitted to the main project branch. Qe mined 18,110 changes from Aug. 09 - May 13. Using an empirically built and validated set of keywords, we identified 741 potential VCCs. Using a two step manual analysis process, we determined that 60 changes were VCCs. Figure 1 overviews our approach.

We hypothesized that: *"developers introduce more vulnerabilities early in a project when it is less mature and developers are less focused on security"*.

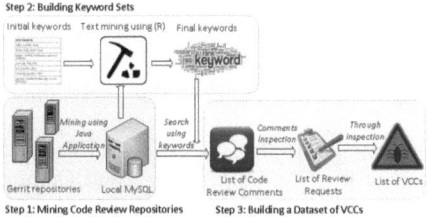

Fig. 1. Research Method

3 Results and Conclusion

The red line in Figure 2 shows that for AOSP, project age does not seem to be the cause of VCCs. Similarly, the gray line in Figure 2 shows that the ratio of VCCs/total changes also does not seem to describe a pattern for introduction of VCCs. As neither of these factors explained the pattern, we conducted exploratory analysis on another factor, release cycle (vertical green lines in Figure 2). It seems that the release cycle may be able to partially explain the introduction of VCCs. The number of VCCs was generally increasing prior to a release and decreasing after a release (although the pattern does not hold in all cases). These results suggest that further study of VCC introduction patterns is needed across more projects. There appear to be factors other than age and total number of changes that may predict when VCCs are introduced. As further research identifies these factors, we can provide more concrete advice to project managers regarding when to emphasize security testing.

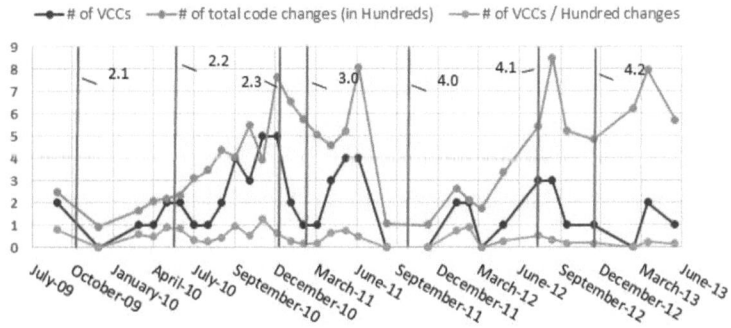

Fig. 2. Number of VCCs identified during code review over 45 months

Acknowledgment. This research is supported by NC State Science of Security lablet.

References

1. Meneely, A., Williams, L.: Secure open source collaboration: an empirical study of linus' law. In: Proc. 16th ACM Conf. on Comp. and Comm. Security, pp. 453–462 (2009)
2. Neuhaus, S., Zimmermann, T., Holler, C., Zeller, A.: Predicting vulnerable software components. In: Proc. 14th ACM Conf. Comp. and Comm. Security, pp. 529–540 (2007)

Author Index

Abad, Abel Meneses   195
Ahmed, Iftekhar   181
Alves, Fernando   191
Ameller, David   168
Armuelles, Iván   111
Armuelles Voinov, Iván   107
Azarbakht, Amir   41

Baiyere, Abayomi   113
Ben-Jacob, Ron   168
Bonnin, Jean-Marie   103
Boodraj, Maheshwar   172
Bosu, Amiangshu   31, 234
Bouabdallah, Ahmed   103
Braud, Arnaud   103

Cabreja, Yordanis   172
Cáceres Navarro, Haniel   218
Carver, Jeffrey C.   31, 234
Castilho, Marcos A.   226
Cedeño, Aidelen Chung   107
Chacon-Rivas, Mario   143
Chua, Bee Bee   70
Chung, Joaquín   107
Cotugno, Franco Raffaele   61

de Almeida, Eduardo Cunha   226
de Bona, Luis C.E.   226
Deroncelé, Edilberto Blez   218

Espino, Juan P.   111

Febles Parker, Michel Evaristo   176, 218
Fernández, Javier   232
Fernández del Monte, Yusleydi   176
Fírvida Donéstevez, Abel Alfonso   147, 212, 218
Franch, Xavier   168
Fromentoux, Gaël   103
Fuentes, Allan Pierra   147, 218
Fuentes Rodríguez, Juan Manuel   212

Gallegos, Mario   1
García Rivas, Dairelys   212, 230
Garita, Cesar   143

Georgiev, Anton B.   93
German, Daniel M.   51
Gerosa, Marco Aurélio   153, 199
Ghorashi, Soroush   181
Goñi, Angel   172
González, Grace   107
González-Barahona, Jesús M.   1, 123
González Muño, Marielis   147

Hafiz, Munawar   234
Hammouda, Imed   21
Hannemann, Anna   11
Hilley, Patrick   234

Inoue, Katsuro   51
Izquierdo-Cortázar, Daniel   1

Janczukowicz, Ewa   103
Janes, Andrea   83
Janni, Derek   234
Jensen, Carlos   41, 133, 181

Kamei, Fernando   202
Kenett, Ron   168
Klamma, Ralf   11
Koo, José M.   111

Leyva Samada, Lisandra Isabel   195
Liiva, Kristjan   11
Lincoln, Max   191

Manabe, Yuki   51
Mancinelli, Fabio   168
Messina, Angelo   61
Montes León, Sergio Raúl   123
Morgan, Becka   133

Noda, Tetsuo   216

Ortiz, Susana Sánchez   76

Pasqualin, Diego   226
Peñalver Romero, Gladys Marsi   195
Pérez Benitez, Alfredo   76
Pérez Quezada, Bladimir Jaime   232
Pinto, Gustavo   202
Possamai, Cleide L.B.   226

Author Index

Rastogi, Ayushi 164
Remencius, Tadas 83
Robles, Gregorio 1, 123
Rosales Rosa, Eugenio 147, 212

Salminen, Joni 80
Sánchez C., Luis E. 123
Siena, Alberto 168
Sillitti, Alberto 83, 93
Silva, Marco Aurélio Graciotto 153
Steinmacher, Igor 153, 199
Succi, Giancarlo 83, 93

Sureka, Ashish 164
Susi, Angelo 168
Syeed, M.M. Mahbubul 21

Tansho, Terutaka 216
Teixeira, Jose 80, 113
Tejera Hernández, Dayana Caridad 218
Todt, Eduardo 226

Villarreal, Rubén 111

Weingaertner, Daniel 226

MIX
Papier aus verantwortungsvollen Quellen
Paper from responsible sources
FSC® C105338

If you have any concerns about our products,
you can contact us on
ProductSafety@springernature.com

In case Publisher is established outside the EU,
the EU authorized representative is:
**Springer Nature Customer Service Center GmbH
Europaplatz 3, 69115 Heidelberg, Germany**

Printed by Libri Plureos GmbH
in Hamburg, Germany